Make It Rain

Make It Rain

STATE CONTROL OF THE ATMOSPHERE
IN TWENTIETH-CENTURY AMERICA

Kristine C. Harper

The University of Chicago Press CHICAGO & LONDON

The University of Chicago Press, Chicago 60637
The University of Chicago Press, Ltd., London
© 2017 by The University of Chicago
Published 2017
Paperback edition 2018
Printed in the United States of America

27 26 25 24 23 22 21 20 19 18   2 3 4 5

ISBN-13: 978-0-226-43723-1 (cloth)
ISBN-13: 978-0-226-59792-8 (paper)
ISBN-13: 978-0-226-43737-8 (ebook)
DOI: https://doi.org/10.7208/chicago/9780226437378.001.0001

Library of Congress Cataloging-in-Publication Data

Names: Harper, Kristine, author.
Title: Make it rain : state control of the atmosphere in twentieth-century America / Kristine C. Harper.
Description: Chicago ; London : The University of Chicago Press, 2017. | Includes bibliographical references and index.
Identifiers: LCCN 2016028694 | ISBN 9780226437231 (cloth : alk. paper) | ISBN 9780226437378 (e-book)
Subjects: LCSH : Weather control—United States—History. | Science and state—United States—History.
Classification: LCC QC928.7 .H37 2017 | DDC 551.680973—dc23 LC record available at https://lccn.loc.gov/2016028694

♾ This paper meets the minimum requirements of ANSI/NISO Z39.48-1992 (Permanence of Paper).

CONTENTS

ACKNOWLEDGMENTS

This book has been helped along by several organizations and a larger number of individuals. I gratefully acknowledge the financial support I received while undertaking the research for and the writing of this monograph from the Dibner Institute for the History of Science and Technology, Cambridge, Massachusetts, which awarded me a postdoctoral fellowship for the 2004–5 academic year; the National Endowment for the Humanities, which awarded me a year-long fellowship (FB-53252-07) during the 2007–8 academic year; and the Tanner Humanities Center, University of Utah, which also awarded me a residential fellowship for the 2007–8 academic year.

I very much appreciate the efforts made by the many archivists and librarians who helped me find and obtain needed materials, including those at the Library of Congress Manuscript Room; the National Archives in Washington, DC, and National Archives II in College Park, Maryland; the Harvard University Archive; the Massachusetts Institute of Technology Archive; the American Heritage Center at the University of Wyoming; the Dwight David Eisenhower Presidential Library; the Lyndon Baines Johnson Presidential Library (especially Allen Fisher); the National Center for Atmospheric Research Archive (Diane Rabson); the National Academy of Sciences Archive (Janice Goldblum); M. E. Grenander Department of Special Collections and Archives, State University of New York, Albany (Geoff Williams); the State of Washington Archive; the University of Washington Archive; the Utah State Archive; the National Oceanic and Atmospheric Administration (NOAA) Central Library, Silver Spring, Maryland (Doria Grimes); the Skeen Library at the New Mexico Institute of Mining and Technology; and the Florida State

University Libraries. A very special thank-you to the family of Captain Howard T. Orville, which so generously shared Captain Orville's personal papers related to his chairmanship of the Advisory Committee on Weather Control in the 1950s, and to Michael Brown, son of George Brown, pilot of one of the Cessnas involved in the secret GROMET seeding project in India in 1967, who reached out to share information about his father's experience. I am also indebted to the late Willard S. Houston, Captain, US Navy, who shared his diaries with me during an oral history interview, thus leading me to the GROMET project. Thanks also to Maureen MacLeod for indexing several stacks of archival documents.

For assistance with the images contained herein, I thank Doria Grimes, who rooted through stacks of photos for me at the NOAA Central Library; Chris Hunter of the Museum of Innovation and Science, Schenectady, New York; Eisha Leigh Neely, Division of Rare and Manuscript Collections, Carl A. Kroch Library, Cornell University; David Delene, editor, *Journal of Weather Modification*, and his assistant, Wanda Seyler; Liz Lyons, Sandra Licata, and Lucinda Whitehorse of the New Mexico Institute of Mining and Technology; and Devin Soper of Florida State University's Strozier Library for technical assistance with permissions.

I also appreciate the insights of those who commented on oral or poster presentations I made at meetings of the American Society for Environmental History; the Society for Historians of Foreign Relations; the Association for Slavic, East European, and Eurasian Studies; the American Geophysical Union; the American Meteorological Society; and the First World Conference of Environmental History, and at the following institutions: Leiden University; Aarhus University; the University of Manchester; Wolfson College, Oxford; the Niels Bohr Institute, Copenhagen; the Royal Institute of Technology, Stockholm; New Mexico Institute of Mining and Technology; Oregon State University; the University of Oregon; and Florida State University's Departments of History, Meteorology, Geography, and Scientific Computing. A special thank-you to my history department colleagues Robin Sellers, who read, proofed, and commented on the entire manuscript *before* it was trimmed down to a manageable number of pages, and Anne Marsh, who solved my last-minute imaging problems. In addition, I also benefited greatly from comments and suggestions received from my colleagues while I was in residence at the Dibner Institute for the History of Science and Technology, Cambridge, Massachusetts, and at the Tanner Humanities Center, University of Utah, particularly center director Robert Goldberg.

At the University of Chicago Press, I am indebted to my editor, Tim Mennel, for keeping me moving in the right direction, and to Rachel Kelly, for answering numerous questions and helping me to get all the details right as the manuscript moved into production. I am particularly indebted to the three anonymous reviewers, whose detailed, spot-on suggestions were critical to the successful completion of this manuscript. In addition, I greatly appreciate Jo Ann Kiser's expert copyediting.

Throughout the research and writing of this book I have received steadfast encouragement and support from my husband and colleague, Ron Doel. He spotted relevant archival material in unexpected places, remembered arcane bits of information from articles I had overlooked, pulled me out of archives when it was time to hunt for food instead of tasty historical tidbits, commented on several drafts of each chapter without a single complaint, and took on the task of cleaning up images. Thank you, Sweetie! Any remaining mistakes are strictly my own.

You have to figure out how to guide that cloud . . .

SENATOR GEORGE SMATHERS (D-Florida), 1951

Weather control. Juxtaposing those two words is enough to raise eyebrows in a world where even the best weather models still fail to nail every forecast and the effects of climate change on sea-level height, seasonal averages of weather phenomena, and biological behavior are being watched with interest, regardless of political or scientific persuasion. But between the late nineteenth century, when the United States first funded an attempt to "shock" rain out of clouds with cleverly deployed explosives, and the late 1940s, rainmaking (as it had been called earlier) became *weather control*. Methods then under development intrigued people in the highest reaches of the American state. Clear fog for landing aircraft? Check. Punch large holes in clouds to allow departing aircraft to leave and returning aircraft to spot their landing field? Check. Gentle rain for strawberries? Sufficient rain in semi-arid regions for grain production? Enhanced snowpacks for hydroelectric utilities? Drying up clouds so that precipitation would not damage ripe cherries? Check, check, check, check. By the late 1950s, proposed uses and the money to undertake them continued to increase. Enhance agriculture at home and dry up the land of our enemies? Trigger precipitation from clouds to put out forest fires? Put an end to "evil" hurricanes? All that and more. Weather control was not a pie-in-the-sky concept. It was a *water*-in-the-sky, sun in the upper Midwest in winter, fog-free runway, and nontraceable weapon of war concept. It was designer weather on demand.

How might one tell a story about the history of weather control, given the surfeit of jaw-dropping pieces of correspondence, newspaper accounts, scientific articles, government reports, and legislative hearings? The challenge is greater than one may at first imagine, for the options are many. I could focus on the engineering behind the seeding devices, and how and why they changed over time—that is, a history of technology. A history of science could examine the theoretical constructs underpinning weather control and various internal and external events shaping their development. A social history of weather control could answer questions like, How did people think about weather and climate, and the ways to adapt to them? Or, How did religious beliefs fit into these ideas? A legal history of weather control is a possibility—there have been enough lawsuits aimed at weather controllers to develop conclusions about how it fits in with other environmental cases; when these were added to laws that were proposed and passed, or proposed and shot down, one could find more than enough material to support such a study. A study of the many ideas related to environmental control? That would be intellectual history. How were weather control ideas and claims treated in newspapers large and small, and in popular literature? A history of popular culture would do the trick. Alternatively, one could connect weather control to geoengineering writ large or to inadvertent weather modification, the approach taken by James Rodger Fleming in *Fixing the Sky*, in which he reminds would-be geoengineers that they are following in a long line of mythical beings, fictional characters, scientists, and engineers who have long sought to change the atmosphere.[1] A military history focused on the tactical and strategic uses of the weather weapon, or a diplomatic history examining the state's use of weather control to win friends and influence allies during the Cold War along the same lines as Atoms for Peace, Food for Peace, and Water for Peace—both are options. Taking a different route, one could narrate the weather control story as a case study examining the role of expertise in making high-level decisions, or as a history of moral decision making, or as a history of conspiracy theories.[2]

But instead of going narrow, I decided to step back and look at the weather control story from as many angles as possible. All those viewpoints led back to the same starting point: weather control in the United States reached dizzying heights in the 1960s because it was heavily funded and adopted as a foreign and domestic policy tool by the federal government, and then plunged into virtual oblivion on the public stage in the 1970s as the funding dried up.[3] Since the whole is greater than the sum of these already intriguing parts, *Make It Rain* thus weaves together the story of how federal officials—politicians, bureaucrats, military leaders, or diplomats—pursued their goal of producing designer weather on demand, even as meteorologists were struggling to figure out how the atmosphere functioned, how to model it, and how to make viable weather and climate forecasts. For while state officials knew what they wanted—weather on demand—the meteorologists knew what they did not know: under what circumstances clouds produced rain, snow, drizzle, or any other hydrometeor, to use the all-purpose scientific term. And no matter how proponents of weather control spun it, weather control was really about controlling only one weather element: water. Not atmospheric circulation, not temperature, not atmospheric pressure. Just water. Water for drought-stricken areas, the US West, and increasingly crowded urban areas. Water for hydroelectric power, for irrigation, for transportation, and for recreation. Water for life for our friends and allies. Too much or no water for our enemies.

What we know now, and what observers could have discerned at the time, was a very large disconnect between weather control advocates and meteorological experts. Decision makers in the early 1950s were being enticed to consider weather control as a real possibility for whatever weather problem ailed them by prominent scientists whose reputations screamed out "elite experts": Nobel Prize–winning chemist Irving Langmuir of the General Electric Research Laboratory in Schenectady, New York; famed computer architect and mathematician John von Neumann of the Institute for Advanced Study at Princeton University; physicist Vladimir Zworykin (known for his invention of the scanning television camera) at the Radio Corporation of America (RCA) Laboratory, also in Princeton, New Jersey; and thermonuclear bomb creator Edward Teller. But while all of these men were convinced that controlling the weather was just a matter of lab time and money, meteorologists—whether members of the US Weather Bureau or academics in research universities—understood that controlled laboratory conditions had little relation to the wild, chaotic atmosphere and its interactions with Earth's varied surface, where one did not get to select the initial conditions before an experiment began nor get to control the many variables that influence what we call weather. Consequently, mid-twentieth-century meteorologists were loath to get involved with anything smacking of weather control other than the microscale controls available to farmers who chose to save their crops from freezing temperatures by smudging, spraying water on plants, or circulating air with large fans. They *knew*, at a time when it was still difficult to make a good forecast thirty-six hours in advance, that controlling the weather over hundreds of acres of grassland or mountain ranges, much less continents, was out of the question. And bringing massive hurricanes and tightly wound tornadoes to heel? As New Yorkers might say: fuhgeddaboudit! But our distinguished nonmeteorologist experts had sufficient clout to make federal leaders take notice and, in turn, dismiss the concerns of atmospheric scientists who were attempting to explain the difficulties involved.

Dazzled by promises of water in the sky becoming "gently falling" water on the ground, members of the US Congress have been holding hearings on Capitol Hill about weather control for more than fifty years even though humans' ability to control the weather—which was "just a few years away" in the 1950s, and then the 1960s, then the 1970s—remains elusive. The reasons behind supporting weather control—drought, floods, famine, wars, fires—change with the decades, but the tantalizing idea of controlling the uncontrollable remains firmly entrenched. Now . . . with the global temperature rising, larger and larger chunks of ice calving from Antarctica's ice sheets, glaciers melting at a faster clip, sea level rising, weather appearing wildly erratic—although when one gets hit by nasty weather it generally seems wildly erratic whether it is within norms or not—people will start itching for their leaders to do something, preferably quickly, painlessly, and cheaply. Such was the reasoning in the late nineteenth century when the first federal appropriations were granted for explosives to shock water out of the sky over the dry Texas plains. It continued through the twentieth century, and it will likely continue through the twenty-first and beyond. And that is why and how "weather control: a grand idea" turned into "weather control: a state tool."

CONSIDERING THE "STATE"

But what do we mean by the state? It seems deceptively easy, yet "speaking coherently about the state," writes sociologist Patrick Carroll, "is far more tricky than it might at first appear."[4] No kidding. And even trickier is doing so without resorting to jargon-laden, head-scratching prose unique to the academic tribes who think and write about the state. Nevertheless, we cannot discuss weather control as a state tool without first engaging with the question, what is the state?

A typical starting point is Max Weber's early twentieth-century argument that a state is a coercive organization that controls territories and those who reside therein.[5] That sounds an awful lot like a "government," but political scientist Alfred Stepan argues, "The state must be considered more than the 'government.' It is the continuous administrative, legal, bureaucratic, and coercive systems that attempt not only to structure relationships *between* civil society and public authority in a polity, but also to structure many crucial relationships within a civil society as well."[6] And political scientist Stephen

Skowronek describes a "sense of the state" in American politics that "is the sense of an organization of coercive power operating beyond our immediate control, and intruding into all aspects of our lives" even though "the absence of the sense of the state . . . has been the great hallmark of American political culture."[7]

The rise of weather control efforts took place against this "absence of the sense of the state" as the United States' bureaucratic underpinnings began to flower during the late nineteenth and early twentieth centuries, helped along by the reform movement of the Progressive Era, by World War I, and by the New Deal Era that accompanied the Great Depression. The further expansion of the federal bureaucracy (also termed "state building") picked up steam during World War II and continued on at full speed during the Cold War, not only due to the expansion of national security concerns but also due to responses to civil rights and environmental concerns that had been simmering below the surface for decades. Throughout the twentieth century, the state bureaucracy had been building upon its capacity to take action through various agencies' efforts to use science and technology to control nature on a variety of fronts via, for example, dam and levee building, modification of river channels, insect control, and forest management in addition to weather control. In all of these instances, we are seeing the federal government bureaucracy expand and "mold institutional capacities" as it responded to domestic and international crises, class conflicts, and the increasing complexity of daily life.[8] Historians, political scientists, sociologists, anthropologists, and other social scientists have used many terms to describe this emerging state, including administrative, bureaucratic, welfare, industrial, post-industrial, capitalist, regulatory, warfare, garrison, national security, proministrative, scientific, technology, and engineering. But the pithiest way to look at those who sought to use weather control harkens back to economist and labor historian John R. Commons's "The state is . . . officials-in-action." Or, as historian Margot Canaday writes, "What officials do."[9] I prefer this definition, particularly for the weather control story, because it is essentially all about officials in action.

Those officials—elected, appointed, civil servants—may only take action to regulate growing seasons or to control the weather if the state has both the autonomy and the capacity to act. Let's consider autonomy first. Sociologist Theda Skocpol argues that in the United States, state autonomy has been historically in short supply because state power is "fragmented, dispersed, and everywhere permeated by organized societal interests." However, she also notes that Stephen Krasner, in *Defending the National Interest*, uses state autonomy to explain US foreign policy, arguing that the president and secretary of state are relatively insulated from "specific societal pressures" and hence act in "furthering the nation's general interests." Political scientist Daniel P. Carpenter argues that bureaucratic autonomy depends on bureaucracies being politically differentiated from those who seek to control them, able to solve problems and create plans, and seen as legitimate organizations.[10]

But aren't all federal agencies tasked with furthering the nation's general interests? And if that is the case, aren't all of these federal officials acting on behalf of the state at least somewhat removed from special interests? Certainly in the case of weather control, agencies that were actively promoting its development (Departments of Defense and Interior) and those that were not (Weather Bureau) were acting under their own volition and thereby autonomous. What of Carpenter's conditions for autonomy in the weather control case? That is a little bit tougher, particularly for the Weather Bureau. It certainly had interests and a worldview that diverged from the politicians who were trying to influence it, but it lacked the power to successfully push back against them. It had the ability to solve problems, if by that we mean the ability to efficiently provide appropriate weather forecasts and warnings throughout the nation. What it didn't have, primarily due to budgetary constraints, was the ability to create new programs. And what it completely lacked was a strong organizational reputation, perhaps because before World War II ended, meteorology wasn't considered much more than a "guessing science."[11] The Departments of Defense and Interior were in much better positions to fulfill the first two conditions, although Interior suffered from organizational problems in the early twentieth century. So while the Weather Bureau, the agency most likely to be involved in weather control, had little autonomy, the military services, backing the initial experiments, had considerably more.

And what of state *capacity?* Skocpol writes that, to be effective, a state needs to have control over its territory, loyal, skilled officials, and adequate financial resources. But even a state that possesses all three of these attributes will not necessarily be able to successfully enact and carry out policies in every area under its jurisdiction. Some of those policies, Skocpol points out, may be stronger or weaker than others.[12] In the US case, we have territorial integrity, the financial means, and the staffing to fulfill the requirement for state capacities. How does that play out for the relevant agencies in the fights, some secret, some in the public view, over implementing weather control? The Weather Bureau, in particular, had an extremely dedicated workforce, which before World War II included almost every meteorologist in the country. Similarly, the military services and the Department of the Interior's Geological Survey had significant numbers of highly qualified scientists. Interior's Bureau of Reclamation, focused on practical undertakings such as building dams, tended to have less scientific strength and significant organizational problems early in the twentieth century, but had significantly enhanced its engineering staff by midcentury. Financially, the Weather Bureau was always in the worst shape, invariably operating on a shoestring budget compared to other units and consistently lacking research funding.[13]

In any event, for our purposes the state encompasses the actions of officials in all branches of the federal government from the president on down, and in all relevant departments, agencies, and bureaus. Some of those actions were taken directly; others were made possible by government contracts with universities and private companies. But the entity exerting control via funding and policy decisions was the American state, which was being legitimized by its "instrumental uses of science and technology."[14]

SCIENCE AND THE STATE

So how are science and the state connected? This, too, seems deceptively easy. Science and the state have been intertwined since sixteenth-century English natural philosopher and courtier Francis Bacon suggested using the power of science to control nature for the benefit of the state, and for the state to fund science to fulfill that purpose. Some scientists think their work is unaffected by anything outside of their laboratories or field sites, whether politics or patronage. But scholars who have examined this relationship present a very different story: science has been political for centuries and is integral to modern states' political systems. Science and the state are about control— the former controls nature and the latter controls territory and people. And since the control of territory also involves controlling the environment in which we all live, the practices of science and the state are effectively joined at the hip. With the rapid technological revolution of the last couple of centuries, science, technology, and the state are so closely intertwined that it is almost impossible to tease them apart.[15] As sociologist Sheila Jasanoff puts it, "there cannot be a proper history of scientific things independent of power and culture. . . . Science and technology operate, in short, as *political agents*."[16]

In *Make It Rain*, I view weather control as a political agent: politicians, with the aid of entrepreneurial scientists, were (and are) attempting to use it for their own political ends, be that domestic (bringing home water to their states) or foreign (as a weapon or diplomatic tool). But those entrepreneurial scientists and their scientific and technological expertise were not playing on the same level as the state officials calling the shots: they were subordinates, taking the funding and producing the science, but not in a position to determine how it would be used.[17] The scientists, who received state funding to conduct research in the many facets of weather control, particularly during the Cold War, may have naively considered their research as basic as opposed to applied, but the eventual uses of their discoveries about atmospheric behavior were not going to be under their control, or even under their influence. Weather control was in the hands of the American state.

Make It Rain unfolds in three parts. Part I, "Weather Control: From Scientific Fringe to Scientific Mainstream," opens the story in 1891 with the earliest federal appropriation for weather control (at the time, most often termed "rainmaking") and its less-than-scientific explosive techniques (chapter 1) to its post–World War II introduction into the scientific mainstream due to research conducted at the General Electric Research Laboratory (chapter 2).

Part II, "Coming to Grips with Weather Control (1950–1957)," takes a look at what happened as weather control took its first tenuous steps out of the laboratory into practical application as seen by federal lawmakers, state lawmakers, and meteorologists—essentially their stories are told independently and then folded together at the end of this section. Chapter 3 examines how federal legislators' belief that weather control would become the equivalent of atomic energy for both domestic and military purposes led them to propose the creation of a massive bureaucracy modeled on the Atomic Energy Commission, which would keep weather control under control. As they saw it, controlling weather control was critical for the nation's defense posture during the Cold War. Failure to impose federal control, moreover, could lead to massive domestic problems (states stealing precipitation from downwind states, floods in some places, droughts in others), and worse, it might allow the Soviet Union to perfect weather control first and thus be able to control the world. Chapter 4 moves away from Washington, DC, to the Desert Southwest and the Pacific Northwest to examine what happened when commercial cloud seeders started selling their "rain enhancement" services across the country, leaving unprepared state governments to sort out the disputes among competing customers of designer weather. Unlike congressmen who anticipated using weather control as a state tool, state legislators and bureaucrats found themselves gobsmacked by piles of irate letters from constituents who were convinced that their excess rain (State of Washington) or lack of rain (states in the Intermountain West and Desert Southwest) was directly due to the actions of "rain enhancers" who operated under the regulatory radar. Exactly how were they to regulate a service when they knew neither how nor when nor where it was taking place, nor if it even worked? How could they keep all their constituents happy when different parts of their states had different concepts of "ideal weather"? Moving from the realm of government to the realm of gun-shy meteorologists, chapter 5 follows the efforts of meteorologists within and outside the United States who were attempting to sort out the weather control "problem." And a problem it was for members of a scientific discipline that had managed to scrape up some professional credibility during World War II and did not intend to lose it as the weather control controversy played out in scientific journals, on the front pages of national newspapers and covers of major popular magazines, and in congressional hearing rooms. Just as meteorologists seldom are in total agreement over a weather forecast, the meteorologists of the 1950s were not in total agreement over weather control. They generally agreed on the need to conduct a lot more basic research on cloud physics and precipitation mechanisms—and were happy to take the proffered funding to do so—but they fell into multiple camps when it came to efforts to clear fog, enhance rain, mitigate hail, snuff out hurricanes, eliminate droughts, and any of the other possible uses of weather control. Commercial meteorologists whose firms were offering weather control services wanted to press forward with practical applications before the basic theoretical work was solidly in place, academic meteorologists generally leaned toward doing the research first and maybe applying what they learned later, and the Weather Bureau meteorologists . . . well they were aghast that anyone would try to make money from the weather, much less promise to solve weather problems with a single pellet of dry ice. Could they work out their differences and still look professional while they were doing so? Would they hunker down in their respective camps waiting for the weather control storm to blow over? Or would they be able to wrest control of their research agenda back from federal lawmakers who knew little about the atmosphere other than they wanted it changed?

Part III, "Weather Control as State Tool (1957–1980)," begins in the late 1950s as President Eisenhower's Advisory Committee on Weather Control submitted its final report just two months after the Soviet Union launched Sputnik and heated up the Cold War's space race. Amid concerns that the Soviets would control the world's weather before the United States, funding for weather control research kicked into high gear, followed by even more funding for weather control applications. Chapter 6 examines efforts to use weather control techniques for domestic purposes by discussing three such cases. The first sought to reduce lightning strikes, thus reducing the number of lightning-caused fires—and if the fires started anyway, to extinguish them with induced precipitation (US Forest Service: Project Skyfire). The second sought to tap water from the atmosphere to fill existing reservoirs that fed irrigation and hydroelectric power systems (Bureau of Reclamation: Project Skywater). And the third sought to snuff out hurricanes while they were small or, alternatively, steer the bigger ones "harmlessly out to sea" (Weather Bureau/Air Force/Navy: Project Stormfury). Skyfire started in the early 1950s, while Skywater and Stormfury began in the 1960s, but by the 1970s they all lost their funding and were discontinued. Why was that? There were a variety of reasons (all explored in the conclusion of Part III), but the fallout from the military's weather control efforts was not helpful. Those efforts are discussed in chapter 7. While the domestic programs were being touted in the press with the help of black-and-white glossy photos and breathless press releases, the military and diplomatic programs were very much SECRET even though the same (military) personnel behind the Project Stormfury hurricane-control project were using classified versions of the same techniques to attempt to secretly break the devastating mid-1960s Bihar drought in India (Project GROMET) and simultaneously to wash out the Ho Chi Minh trail and North Vietnamese military emplacements during the Vietnam War (Project Popeye and Project Compatriot). Although the story of weather control efforts in Vietnam was broken in the *Pentagon Papers*, detailed by columnist Jack Anderson and journalist Seymour Hersh in the early 1970s and then disclosed more fully in hearings led by Senator Claiborne Pell (D-Rhode Island), much of the material on both the India and Vietnam weather control efforts has only been declassified in the past ten to fifteen years. I examine what members of the Lyndon B. Johnson administration were thinking when they ordered the execution of these projects, and how using weather as a weapon jibed with using weather to improve agricultural output and keep the home front strong. The Conclusion draws the story to a close, examining why weather control faded as a state tool in the 1980s, but never really died completely . . . and how it may return as an "answer" to the weather that ails us and the problems with the natural resource none of us can live without: fresh water.

* I *

Weather Control: Scientific Fringe to Scientific Mainstream (1890–1950)

> There can be no full conquest of the earth, and no real satisfaction to humanity, if large portions of the earth remain beyond his highest control.
>
> JOHN WIDTSOE, 1928[1]

CONTROL (WHAT IT'S ALL ABOUT)

Before diving into weather control in its early guises, let's talk about control in general. As we have seen, by definition, the state controls territory and the people living within it. Expanding territory and population requires ever more complex control over time.

With its ofttime fellow traveler, technology, science is about control as well: the control of nature. And if the state and its people can use science and technology to control nature, then why not control the *weather* and thereby the production and distribution of water for the nation's benefit? During the late nineteenth century, the United States was overrun by technological enthusiasm. Professional engineers and tinkerers alike were masters of innovation and invention, creating and producing new communications devices (telegraph, telephone); more efficient steam engines to power railroads, ships, and factories; and internal combustion engines that would power automobiles and, in a few short years, airplanes. These innovations brought people closer together as travel and communication became faster and easier.

Writing in *Scientific American* (1896), Edward W. Byrn called the patent-rich period following the Civil War "an epoch of invention and progress unique in the history of the world . . . a gigantic tidal wave of human ingenuity and resource."[2] Thomas Parke Hughes, the historian of technology, later agreed: "Interest in invention and inventors was a manifestation of the realization of the power of technology."[3] By the end of the century, many Americans thought technology "was a broader, generalized, man-made force that could be applied at will to a wide variety of problems as they arose. Technology could bring order out of chaos, provide boundless energy, support business enterprise, and win wars."[4] And so the idea of controlling the weather was not really out of touch with the times—despite a lack of underlying scientific theory. Indeed, it made perfect sense, taking its place among all of the other rational, efficient methods that were being used or being proposed to be used to control forests, fisheries, agricultural output, or water resources.

THE PROGRESSIVE ERA: PUTTING SCIENCE TO WORK FOR THE STATE

The market was driving innovation in the late nineteenth and early twentieth centuries, and industry was striving to meet the demand.[5] The combined synergistic effect of science-supporting universities, nascent scientific professionalism, early corporate research and development, the introduction of scientific elites to the political process, and the rise of philanthropic foundations that were funding scientific efforts heightened the success of all of these entities.[6] As the state grew in tandem with the professionalization of science, academics began offering solutions to state problems. What the state did not do was provide funding for these scientific efforts. In the days of "small science," philanthropic donations were sufficient to keep laboratory and fieldwork going. Indeed, as scientists were attempting to find solutions for societal problems, they took great pains to maintain their objective and disinterested status by not seeking federal funding. The role of federal funding for science would not come into play until after World War II.[7]

The drive for innovation was already well underway when the scientifically informed, reform-minded responses to Gilded Age excesses became focused during the Progressive Era (1890 to 1920, give or take a few years). In their "search for order," as historian Robert Wiebe put it, reformers wanted to apply rational, that is, scientific and efficient, controls to the workings of government that would encourage more democratic participation while also putting a premium on the use of experts to find solutions to major problems.

At the same time, waves of new immigrants were pouring in from central and southern Europe, many of them Roman Catholic, Jewish, and peasants. But instead of walking off ships at Ellis Island and heading west to establish farms as earlier immigrants had done, these new arrivals settled in East Coast cities where they found opportunities to make their way in a new land. Faced with rapid population growth, big city governments had to provide them with basic services, including education, sanitation, transportation, safety, and housing, and jobs as they became part of the fabric of their municipalities. Bureaucracies grew to accommodate those needs, strengthening the state apparatus at all levels.[8]

Anti-urban ideas intensified as the population of dark-haired, dark-eyed, darker-skinned, non-Protestant people grew along with suffering and distress in big cities. Some politicians suggested mitigating this "immigration problem" by packing these folks up and sending them out West. Certainly there was plenty of land for homesteading, as earlier settlers had done when "out West" meant Ohio, not some dry, treeless expanse west of the Mississippi River. But it was dry out there, so expensive irrigation projects would be needed to extract value from the land. Considering that a worldwide agricultural depression was underway at the same time, investing huge amounts of money in irrigation projects did not make a lot of sense. What if, however, there were another way to bring water to the parched land? Enter rainmaking. New inventions were appearing every day, and there was definitely a need for cheap water. So why not pursue it? As journalist Walter Lippmann put it, "We shall use all science as a tool and a weapon."[9] Using science would bring progress to the twentieth-century American state.[10]

Similarly, state bureaucracies grew to provide rational control over natural resources. Gilded Age entrepreneurs had systematically exploited natural resources, including timber, water, minerals, and agricultural land, to advance their industrial and economic agendas. But the devastation they left behind fed fears of scarcity and deprivation, particularly in view of the millions of newcomers whose needs had to be met. In response, society moved from a position of wastefulness to one of centralized efficiency, and conservation took on new importance. Mind you, the conservation of the early twentieth century was not the preservation of wild lands that we think of today when discussing conservation, but the maximum sustainable use of the resource in question. To determine just how to get the maximum use from a resource without depleting it for future generations would require the input of scientific experts, who were standing by to provide it to the increasingly strong bureaucratic state.[11]

Bureaucratic development was most effective in the hands of strong agency heads, who assembled outstanding talent and built legitimacy, and thus their organizations' reputations, as they capitalized on the authority accorded to scientific experts. The agencies that pulled off this feat—the US Department of Agriculture is the premier exemplar—were able to make social changes through the efforts of their in-house experts.[12]

DEVELOPING EXPERTS AND EXPERTISE

In the late nineteenth and early twentieth centuries, scientific authority carried weight with the broader society, and its status remained undisputed until the 1950s.[13] Liberal writers Herbert Croly and Walter Lippmann associated scientific expertise and professionalism with being objective and disinterested, and hence scientists undertook their work in the best interests of both the state and its people. As esteemed solar astronomer George Ellery Hale argued in a 1923 National Research Council report, science was about truth and progress, and "its work for humanity has only just begun." For Hale, his colleagues, and most middle-class Americans, science was "cumulative and ever progressing."[14]

Hence, scientific experts were the ones best suited to solve complex problems for the state, and their authority as scientists was used to obtain administrative autonomy for the agencies that employed them. In addition, federal administrators often worked with professional associations on problems so that it did not appear as if the state were meddling directly in scientific issues. All of these experts came out of academic settings since for many disciplines basic research was undertaken in the new research universities.[15] After World War I, relatively new scientific fields, including meteorology and geophysics, formed professional associations for the first time, and the interwar years proved important for professionalization in those and many other scientific disciplines.[16] Indeed, this professionalization of the sciences fed into the technocracy movement, which called for the institutionalization of technological change for state purposes and argued that civilization's progress was directly tied to scientific progress, and took root during the 1920s.[17] However, elite scientific researchers, that is, those working at prestigious research universities, who were more than happy to weigh in on federal policy issues, avoided federal funding throughout the 1930s because they did not want the state to control their research agendas.[18] They spent the decade creating institutional ties among science, universities, industry, and the state, which were then set into motion by World War II and its massive technoscience undertakings to create all kinds of military hardware, medical breakthroughs, and, of course, the atomic bomb. By the time the war was over, state funding of science was solidly in place, and scientific experts were firmly in service to the state.

ADVOCATING CONSERVATION, BUILDING
BUREAUCRACIES: WASTE NOT, WANT NOT

And advance the ideas behind conservation they did, with a laser-like focus on scientific management. Conservation ideals arose from implications of science and technology in modern society, with professionals and experts using the results of applied science to provide input for federal policy decisions. State-sponsored science and scientists within cabinet agencies were solidly in place before World War I. As might be expected, the federal government was strongly involved with supporting agriculture, and the US Department of Agriculture (USDA) became the premier scientific executive department, with research arms that extended throughout the country. Its subordinate services and bureaus were science-based, with the Forestry Service, for example, borrowing and implementing European techniques of forest management starting in the late nineteenth century, while other USDA offices started promoting insect and weed control to improve crop yields in the first half of the twentieth century.[19]

One of the USDA's subordinate organizations, however, was not as scientifically solid: the US Weather Bureau (USWB). Established under the USDA in 1891, it consolidated earlier federal weather services provided by the US Army Signal Service. But unlike the USDA offices that dealt with agricultural sciences—which were staffed with personnel holding discipline-appropriate college degrees—USWB offices were filled by people who had learned weather forecasting on the job, often starting as observers when they were teenagers and then working their way up to higher positions. Why the difference? US colleges did not offer degrees in meteorology until the late 1920s when the Massachusetts Institute of Technology set up a graduate program within its aeronautical engineering degree program to meet the requirements of US Navy officers needing advanced training. Indeed, the very idea of someone getting a degree in meteorology was something of a nonstarter. As Harvard climatologist Robert DeCourcy Ward put it, "Everyone thinks they are a meteorologist." Why study something that you already know? Unlike in the agricultural field, there were no "expert" meteorologists. Consequently, any meteorological idea was seen to be just as good as any other. So while the Weather Bureau tried to provide services to keep people safe, it had very little credibility. It was, one physicist opined, a "guessing science."[20] And in the Progressive Era, that was not good enough. Expertise depended upon solid science, and meteorology was anything but solid. (We'll return to the theme of expertise—who is an expert and who is not—later.)

Somewhat surprisingly, given John Wesley Powell's famed report on the arid lands of the American West (completed in 1888 by the US Department of the Interior's Geological Survey), Interior was lacking in scientific capacity. In 1888, the USGS had measured water supplies, located sites for reservoirs and canals, and mapped areas suitable for irrigation in the West. The federal government took that information and planned out a strategy for western irrigation to bring water to these dry lands because its presence was critical to the region's development and political economy, and westerners were interested in finding ways to access untapped water resources.[21] By the late 1890s, federal reclamation presented the United States with the ability to engineer its way out of its water-related problems and was seen as a way to aid established western farmers and landowners, as well as to boost private development.[22] Irrigation, in particular, was important to preserving the American way of life, and consequently, as historian Donald J. Pisani notes, "nostalgia for rural America helped make irrigation a popular science," especially as immigrants continued to cram into eastern cities.[23] When the Reclamation Service was founded under Interior in 1902, it appeared that every state could benefit from irrigation. Thus, Reclamation "began its life with a great deal of autonomy and forfeited its ability to engage in either science or comprehensive planning."[24] But what I think Pisani misses here is that hydrology was not an organized science in 1902—indeed trying to nail down what the science of hydrology is today remains difficult—so the Reclamation Service would have found it difficult to embrace a more scientific mode of operation. Instead, it relied on engineering, which was also not exactly "scientific" in the early twentieth century.

The Reclamation Service was on the lookout for "excess water" in the West—that is, water that was not spoken for by anyone under the various state water rights laws, so that it could be stored for eventual use instead of being "wasted" by running to the ocean.[25] Water would also be wasted if it soaked into the ground. The point was to make sure that every drop of water was put to "good use."[26] And other agencies followed this same mantra: control over the resource meant that nature would be conquered and resources would not be wasted. Thus we see state efforts to control rivers via dams, channels, and levees built by the US Army Corps of Engineers.[27] Similarly, the Forest Service under German-trained forester Gifford Pinchot, who used his personal wealth to create a network of forestry experts and founded the Yale School of Forestry because of the importance of professional expertise, was focused on "improving nature" by making "the forest more productive, eliminating waste by cutting old growth and simplifying forest structure" to make it easier to manage.[28] The conservation of forests was critical to getting the maximum possible use from available timber. The idea of "sustained yield," which later became "maximum sustainable yield," would guide not only forest practices but also those related to the game within the forests and the grazing lands that surrounded them.[29] Forestry experts sought to reduce waste in forests by reshaping them in an efficient way. And because timber was so important during times of war, forest conservation was tied to national security.[30] Virgin forests, with their stands of huge old-growth trees, were perceived as reservoirs of wood (much like clouds would become airborne water reservoirs in the 1950s). The only problem? Scientific forestry was easier to undertake in the laboratory, where all variables could be controlled, than in the forest, where they could not.[31]

Aldo Leopold, whose career started in forestry before he turned his attention to game management, sought to use the ideas of sustained yield for game animals as well as for forests. For him, historian Nancy Langston writes, "killing predators meant eliminating waste—a goal at the very heart of conservation."[32] Once the predators were gone, people would be able to hunt and eat the game animals. Otherwise, the predators would eat them, and if no one were interested in eating the predators, then they were using up resources that could have gone to some useful species instead.

The state also controlled fish: the industrial management of fisheries took hold in the early twentieth century as well.[33] And it mobilized against insects that were attacking agricultural lands.[34] State involvement in agriculture—soil science, fertilizers, hybrids, planting and harvesting technologies—was huge, especially during the Great Depression, when unemployment was a problem, and during the Cold War, due to national security concerns.[35] In the late 1940s, the USDA and the Army worked together to make herbicides as inexpensive as possible. This state-led attack on weeds, which was justified by war efforts, encouraged farmers to use herbicides instead of time-honored mechanical methods to keep weeds under control.[36]

BUILDING SCIENCE AND THE STATE

Part I extends from the Progressive Era to World War I, through the Roaring Twenties and the Great Depression of the 1930s, and ends in the immediate post–World War II years. From the very beginning, science and technology were both embedded in and growing with the American state. Starting in the late nineteenth century, science and technology were called upon to solve increasingly complex societal and natural resource problems, and associated experts were tapped to use their expertise for the state. By World War I, physicists and chemists were working on weapons development and deployment for the military, and throughout the 1920s and 1930s, agriculture experts helped the state to gain control over farm output, while engineers controlled water flow and water resources, particularly in the US West. Mobilization for World War II brought tens of thousands of scientists and engineers into state service, and many of them continued that service during the subsequent Cold War. It is into this interwoven fabric of science, technology, and the state that weather control first emerged. At the leading edge, it was one idea among many to bring water to dry western lands and thus revive the American love affair with an earlier agrarian ideal. By the late 1940s, it appeared to be backed by solid scientific expertise. Why and how would the state use it then?

CHAPTER 1

Ka-Boom!

Invention has almost placed the word "impossible" on the retired list.

TAGLINE, *The Inventive Age*, 1891

Three p.m., August 18, 1891, near Midland, Texas. Robert G. Dyrenforth, US Patent Office assistant commissioner and self-titled "General," and a team of assistants exploded a hydrogen-filled balloon some 6,600 feet above the dry prairie, as nonprecipitating clouds scudded past. Cattle moseying nearby watched the preparations with narrowed eyes and raised eyebrows, and promptly stampeded in the opposite direction. Shortly thereafter, electrical charges ignited several dynamite-laden kites. As the air shuddered with the final explosion, the *Washington Post* reported, the air pressure dropped and the "rain came down in 'torrents.'"[1]

Thus began the US government's first foray into weather control, an effort devoid of meteorological theory, scientific thought and methodology, and expert advice. Today the effects of climate change are no longer in scientific dispute as popular media and academic presses alike routinely report on sea-level rise, retreating glaciers, and extreme weather events. Nations and their people are increasingly concerned about the availability of sufficient fresh water from rain- and snowfall. Inventive, people have always sought ways to mitigate and/or adapt to environmental impacts, whether by devising and making shelters and clothing or by damming streams to create a more stable water source for direct consumption or irrigation. But available technologies could not, until recently, change the climate. Today those technologies, some of them more plausible than others, do exist. Should we use them or not? What might happen if we do? What might happen if we don't? Who will get to make those choices?

We have become accustomed to controlling much of nature, whether by planting crops, draining swamps, carving through mountains to ease transportation, using pesticides to kill insects or fertilizers to promote plant growth, or aiming to get the maximum sustainable yield from timberlands or ocean-going fish stocks. In fact, we control so much of nature that we don't often think about it. But while the *idea* of being in control has been in place for centuries—don't we all like to think we are in control most of the time?—the technological ability to be in control has not. And in the United States, a young, by European standards, nation established with a deliberately fragmented government lacking an extensive bureaucratic structure, state control only became a possibility in the late nineteenth century. It came at a time when science and technology were providing opportunities to maintain greater control over our natural surroundings as well as our hemisphere. As the necessary bureaucracies for exerting state control over the entire country were formed, not much seemed impossible. And if people could settle dry lands west of the Mississippi, working the land, channeling water to support agriculture, trying to turn it into some semblance of the land they had left behind, how far-fetched would it be to control the weather? It was just another opportunity to engineer the way out of a problem.

Originally, state control of the weather was about agriculture and the expansion of the West. But within a couple of decades, controlling the weather would be about expanding military aviation, and as World War II wound down and the Cold War spun up, it would be about preserving national security. Weather control, like land surveys, timber cruises, pesticide development, and road building, was a tool in the hands of the state.[2]

As nineteenth-century Americans packed up and moved west to the Great Plains and beyond, they encountered unfamiliar landscapes and climates. In lieu of abundant moisture, flora, and fauna, they found arid and semi-arid conditions supporting grasslands at best, and large tracts of only the heartiest scrub at worst. For settlers moving west of the Mississippi River, change was in order. They could either change the way they lived or they could change the existing environment to resemble what they had left behind, which they did by undertaking relatively small-scale irrigation projects, as Mark Fiege argues in *Irrigated Eden*.[3] Alternatively, they could go big and change the weather, and eventually the climate. In an age of technological enthusiasm, why not? Viable ideas at the time included deliberate weather control by firing cannons into the air or "natural weather control" due to the spread of railroads and farms across the country, as in the "rain follows the plow."[4]

In the nineteenth century, scientific understanding of weather and climate was sketchy by today's standards. How could man or state justify attempts to control the weather? Were allusions to science just a gloss to make them respectable? Or did practitioners of "rainmaking" have a scientific plan?

SCIENTIFIC THOUGHTS ON RAINMAKING— NINETEENTH CENTURY

Following the first state-sponsored rainmaking experiments in 1891, several scientists jumped into the fray, explaining the primary "theories" behind them and recounting earlier attempts to induce rainfall. The essence of their explanations: nothing new here!

Harvard climatologist Robert DeCourcy Ward discussed the possible roles of explosions and fire, reaching back to Plutarch's *Life of Marius* (first century, CE), which claimed that rain fell after battles. Explosives were not in use then, but that did not stop advocates from touting their possible use to trigger rain. By the late eighteenth and early nineteenth centuries, a few scientists had examined and discarded the possibility of using cannons to stop hailstorms and artillery barrages to dissipate clouds and stop thunderstorms. Until the early nineteenth century people were more likely to think that large explosions or other big noise producers actually eliminated storms, but by the mid-nineteenth century this idea had been turned on its head as the idea that large explosions triggered rainfall became more commonly accepted. Puzzled by this 180-degree conceptual change, Australian astronomer and meteorologist Henry Chamberlain Russell surmised that obtaining rainwater had become more important than keeping storms at bay.[5]

Meteorologist James Pollard Espy, self-described "meteorological advisor to Congress," was the first American to propose testing the effects of fires on rain, another old idea. His greatest theoretical work addressed atmospheric moisture and the idea that as rising air cools, the moisture it carries condenses. In midcentury, Espy argued that air moves inward toward areas of rain and "of course upwards," forming large clouds that move, gather additional moisture, and expand. His conclusion: volcanic explosions and large fires could produce rain. If one observed smoke rising from fires, it appeared that smoke contributed to cloud formation, as did particulate matter emitted by factories and chimneys. However, not all fires would produce clouds and rain. Air that held insufficient moisture or was thoroughly mixed by upper-level winds would not be affected by rising smoke. Espy was willing to risk having his experiments fail "if Congress or the State Legislature [would] promise a sufficient reward in case of success." Counting on sufficient summer humidity, Espy planned to assemble large piles of combustible material and ignite them at "various places at once" under favorable atmospheric conditions. Pointing to letters supporting the idea that fires could induce rain, he advised farmers to save brush and timber waste for summer's first dry period and, in a collaborative effort, light fires with others in the vicinity. According to Espy, the farmers would reap a twofer: extended rains benefiting agriculture and brush disposal.[6]

Smithsonian Institution director Joseph Henry was skeptical. Professing respect for Espy's "scientific character" and acknowledging that heat rising above a fire might trigger a storm by overturning air in an unstable atmosphere, Henry did not find the plan economically viable.[7] Congress wisely declined to fund Espy's experiment.[8] But with this storms-from-fires idea on the backburner, using concussions from artillery fire or ringing church bells to induce rainfall periodically resurfaced.[9]

The early 1870s witnessed renewed interest in rainmaking. Writing in *Nature*, mathematician/meteorologist turned naval historian John Knox Laughton acknowledged that Espy had given scientific credence to the old idea of large fires triggering rain. However, noting many more cases where fires had not triggered rainstorms, Laughton argued for care when making cause-and-effect determinations without considering atmospheric conditions. If Espy were correct, wouldn't one expect more rainfall in chimney-filled towns and cities? And yet there was no evidence that London—full of smoke-belching chimneys—got more rain than outlying areas.[10]

Concussions as rain producers suffered similar problems. Theoretically, Laughton wrote, it might be possible to argue that a "violent shock" could cause moisture particles to condense into drops large enough to fall as rain. But where was the supporting evidence? *Whitaker's Almanac* (1869) had mentioned several powder-mill and colliery explosions, but not one had been followed by rain. Similarly, some battles were followed by rain, and many were not. Laughton noted that several major battles had been fought in fine weather that continued after the guns fell silent. He concluded that since storms' causes were still unknown, it was insufficient to point to explosive-caused air movement and fires as proximate causes of precipitation in the absence of favorable atmospheric conditions.[11]

NO EVIDENCE? NO PROBLEM! LET'S TEST!

The lack of causal evidence did not discourage American civil engineer Edward Powers, who published his first edition of *War and the Weather* (1871) in Chicago just a few months before the Great Fire erupted, destroying the published copies and the printing plates. In the years that followed, he traveled the country giving lectures about the possibility of influencing the weather to benefit "the human race and more especially the farmers of America." In so doing, he fit right in with other late nineteenth-century inventors who were developing all manner of contraptions that had the potential to radically change society.[12]

Powers's Espy-influenced book provided evidence based on vague, cherry-picked examples that stated neither the interval between the end of battle and the beginning of rainfall nor whether the rainfall was proportional to the number of artillery fired. Powers neglected to report if the battles had taken place in advance of an approaching frontal system, which would explain the rainfall.[13] These wishy-washy examples, however, held a magnetic attraction for those seeking to change the weather, and his arguments in *War and the Weather* came to provide the underlying support for conducting rainmaking experiments despite a complete lack of support from meteorologists.

Scientific reviews of Powers's book were not favorable. Skewering Powers's argument connecting battles with rainfall as "lame," a reviewer writing in *Silliman's Journal* noted that all of the exemplar battles had taken place in regions where rain typically falls once every three days or so. A battle that started after rain had stopped and lasted for a day or two would almost certainly be followed by rain. Because he failed to provide details about pre-existing atmospheric conditions and any evidence that the rain arrived earlier than it would have without the battle, Powers had, the reviewer wrote, failed to make his case. Powers's failure to provide any examples that did not bolster his case also undermined his argument. In a final slam, the reviewer requested that Powers or someone else discuss this subject using a "truly scientific method."[14]

The Great Fire of Chicago spurred new interest in the effects of fire on rainfall while simultaneously casting doubt on them. Many people thought the fire had been extinguished by fire-induced rain. However, when meteorologist I. A. Lapham, assistant to the chief signal officer (the Army Signal Corps was the nation's weather service until 1891), analyzed the fire and its aftermath, he debunked that idea. Apparently a telegram sent to London had stated the fire was "checked" on the third or fourth day by a heavy downpour, possibly due to fire-induced atmospheric disturbances. Lapham discovered that there had been no downpours during the fire—only widespread gentle rain amounting to a few hundredths of an inch. The fire had died out after exhausting all combustible materials upon reaching the city's extreme northern boundary and Lake Michigan's shore to the east. The downpour? It had occurred four days after the fire was nothing more than smoldering embers. Lapham did not claim, however, that this single example disproved Espy's theory of fire-induced rain.[15]

Twenty years passed between the first and the revised edition of *War and the Weather*, a time in which, Powers noted in the latter edition, extensive efforts to develop "schemes of irrigation" had been underway to supply water in the US West. He scoffed at the idea that irrigation could supply sufficient water to this huge geographical area, and he called attention to the promise of human intervention in the weather process, which depended upon testing the premise that heavy artillery discharges brought rain. Congress, Powers argued, should take legislative action to fund relevant experiments.[16]

Those efforts to develop "schemes of irrigation" stemmed directly from John Wesley Powell's "Report on the Lands of the Arid Region of the United States." Powell, who led several surveying expeditions to the West starting in 1867, argued that the Rocky Mountain Region (which he called the "Arid Region") constituted one-fourth of the nation's land, received insufficient rainfall for agriculture, and would only become productive with an irrigation program, albeit a more limited one than has come to pass. According to Powell, only small areas of the US West that were located along streams were irrigable.[17] Powers wanted to cover the rest with artificial rain.

The "science" behind Powers's arguments was based on rain formation and moisture source concepts found in oceanographer Matthew Fontaine Maury's *Physical Geography of the Sea* and Yale science professor Benjamin Silliman's *Principles of Physics or Natural Philosophy*. Powers argued that streams of moisture flowed through the atmosphere above the United States as the prevailing westerly winds carried water vapor originating from the Pacific Ocean. People needed to tap those streams to provide precipitation to dry areas when nature failed to "act at the proper times"[18]—an argument resurrected in the early 1960s by the Bureau of Reclamation for its Project Skywater.

Powers backed up his "rain follows battles" claim by using examples found in US Navy logbooks and Army officers' accounts of battles in the American Civil War and the Mexican-American War. Anticipating those dismissive of his anecdotal evidence, Powers issued a pre-emptive strike: those unconvinced by his battle accounts would not be convinced by additional examples. So why bother? As he discussed individual battles, he criticized those who took issue with his "science."[19]

Testing Powers's ideas would be expensive. He estimated *each* experiment would cost about $80,000 ($2 million in 2015 dollars), and he wanted to devise and run two experiments: one to originate a storm in the absence of a Signal

Service forecast of an incoming storm, and one to determine if a "Signal Service storm" could be forced from its natural course. According to Powers, after the successful completion of the preliminary experiments, the cost to generate one storm would be about $20,000 ($500,000 in 2015 dollars)—a bargain considering the millions of dollars worth of water that would fall from the sky.[20] Pennies from heaven? Artificial rainmaking was going to be a much bigger deal than that, and one US senator decided to help.

THE FIRST FEDERALLY SPONSORED EXPERIMENTS

In late 1890, Senator Charles B. Farwell (R-Illinois) decided that Congress needed to take a hand in advancing rainmaking. Aware of those rivers of moisture flowing through the atmosphere, he told *Scientific American* that "by means of a sufficient number of first class bangings," the moisture would condense and fall as rain. "It was," according to Farwell, "a question of applying what you know." That scientists knew little and Powers knew less bothered him not one whit.

Farwell's technological enthusiasm was based on the "fact" that heavy rain had fallen after all the great battles of the nineteenth century. He trotted out reports by Senate colleague Leland Stanford (R-California), a Central Pacific Railway magnate whose dynamite-wielding crews had observed rain fall every day in semi-arid areas. Should Stanford's words not seal the deal, Farwell pointed to observations made by German naturalist Alexander von Humboldt, who noted that South American volcanic eruptions in dry seasons fostered rainy seasons and that there were reports of extensive fires in Nova Scotia that had been followed by flooding rains. That settled the matter for Farwell: exploding dynamite triggered rain. He suggested conducting experiments in eastern Iowa, Colorado, or western Kansas, preferably along railway lines, for seven or eight hours a day.[21]

Not so fast, a US War Department signal officer told *Scientific American*. He argued that it would be wise to assess existing atmospheric conditions before launching an explosives barrage to make rain. Setting off ordnance under low moisture conditions—high pressure or just after a frontal passage—was unlikely to produce rain. A better test would involve conducting experiments three hundred to six hundred miles southeast of a low pressure center, placing observers every ten miles or so east of the explosions, up to a distance of two hundred miles. The observers would watch the clear skies for cloud development. However, even if rain developed, he thought it unlikely that its value would exceed the explosives' cost.[22] Perhaps they expected a group of weather enthusiasts to be willing to camp out in the middle of nowhere as they watched for clouds as no-cost, trusted observers.

With abundant, free, just-waiting-to-be-tapped moisture floating overhead, Farwell asked Senate Appropriations Committee members to add $10,000 to the House Appropriations bill "for rain." Laughing, they added his request as a personal favor. *Plus ça change, plus c'est la même chose.* The House removed the "rain appropriation" from the bill; the Senate restored it. With the bill's items listed by number, no one from the House questioned the small amount for "No. 17" and it passed.[23] The appropriation finally totaled $2,000 (about $50,000 in 2015 dollars), later increased to $7,000, and it was given to the Agriculture Department's Forestry Division for experiments using dynamite to induce rain. Why Forestry? Presumably its scientifically inclined personnel would be delighted to find a way to increase rain on woodlands, thus increasing timber yields.[24] Farwell joined Agriculture Secretary J. M. Rusk and several others to consult Patent Office chemist Claude O. Rosell and General Robert G. Dyrenforth, who opened this chapter, on the most effective delivery method for the explosives. They suggested launching balloons containing explosive gas instead of filling the balloons with expensive helium and then weighing them down with explosives. Initially, Dyrenforth had considered using artillery, but changed his mind after getting results from early versions of the oxohydrogen-filled (i.e., a combination of oxygen and hydrogen) balloons tested near Washington, DC. Those tests had shown that exploding a ten-foot-diameter balloon filled with a 2:1 mixture of hydrogen and oxygen produced a very large concussion. Skeptical members of the science-savvy and highly reputable Forestry Division declined to be involved with Farwell's brainchild; Dyrenforth eagerly took charge.[25]

In late summer 1891, Dyrenforth took the train to Midland, Texas, and then bounced in wagons across the Texas prairie to the experimental site on the "C" Ranch, about twenty-five miles away. Joining him in Texas were Rosell, Edward Powers, meteorologist George E. Curtis of the Smithsonian Institution, and Oberlin College's John T. Ellis. The remote location, required for safety, made it difficult to get supplies on short notice. Steady high winds interfered with balloon launches, and the highly alkaline water sickened most of the workers. Nevertheless, they filled the balloons with gas and ignited them with battery-provided electrical charges, as illustrated in the cheeky cartoon in figure 1.1.

Claiming great success—more rainfall over a larger area than in the previous three years—Dyrenforth concluded that the concussions triggered the rain

FIGURE 1.1. General Dyrenforth says to the "professors," "Hurry up the inflation, touch off the bombs, send up the kites, let go the rackarock; here's a telegram announcing a storm. If we don't hurry, it will be on us before we raise our racket." Cartoon by H. Mayer. *The Farm Implement News* 12, no. 11 (1891).

by disturbing upper air currents and "jarring the particles of moisture which [hung] in suspension in the air." Alternatively, frictional (i.e., static) electricity generated by the concussions and the mingling air might have produced, he wrote, a "polarized condition of the earth and the air, and so [created] a magnetic field which may assist in gathering and so condensing the moisture of the surrounding atmosphere."[26] Fascinating! The enthusiastic Dyrenforth was ready to take his successful blasting technique to El Paso, Texas, for another demonstration.

A number of scientists—including the Smithsonian's Curtis, who had an entirely different take on the Midland experiments—were not impressed. Curtis reported that the initial explosives were set off while he was en route to the site. Dyrenforth had reported good but unmeasured "grass rain," the same types of showers Curtis had encountered while approaching from the east. The experimenters, however, did not claim their explosions had triggered the rainfall. Between August 16 and August 20, the team touched off explosives as threatening clouds hovered nearby and Curtis captured 0.02 inch of rain in his gauge, not exactly the "torrents" reported by Dyrenforth.[27]

The evening of the twenty-first, Dyrenforth's team exploded 156 pounds of rackarock (an explosive consisting of potassium chlorate and mononitrobenzene) in fourteen blasts just as a "norther"—a fast-moving cold front that brings precipitation followed by lower temperatures and clear skies—entered the area. Sure enough, the pressure rose, the temperature fell, and a fine mist lingered into the following day. The team connected the mist to the explosions, but Curtis scoffed that the norther had been en route for several days and the mist resulted from incoming cold air wedging up warmer surface air. The last, and largest, experiment took place on the twenty-fifth, under favorable conditions. The explosions stopped at 11 p.m., rolling thunder arrived at 3 a.m., and "torrents" (the favorite adjective) of rain lasted until 8 a.m. However, no one measured the fallen rain, and observers told Curtis that it was "nothing but a sprinkle." Indeed, the weather forecast had called for showers.[28]

SCIENTISTS PARSE THE RESULTS

"In view of these facts," Curtis wrote in *Nature*, "it is scarcely necessary for me to state that these experiments have not afforded any scientific standing to the theory that rainstorms can be produced by concussions." However, he also fretted that some would say that since they did not get definite results, the experiments didn't prove or disprove the efficacy of concussions on rainfall.[29] The federal appropriation had lent credibility to the idea that concussions could lead to rain. Curtis pointed to a *New York Times* editorial that commented, "This theory is really so important that it ought to be thoroughly tested," as it urged additional rainmaking approaches. But, he argued, a congressional appropriation did not make the "theory" more important "scientifically or practically" than it had ever been.[30] Artificial rainstorms, Curtis stated with scientific confidence, would never be triggered by noise or concussions.[31]

Curtis was not the only skeptic. Simon Newcomb, US Naval Observatory astronomer, National Academy of Sciences member, and former president of the American Association for the Advancement of Science, noted that with the success of steam shipping, the laying of the trans-Atlantic cable, and other technological developments, the word "impossible" seemed to be disappearing from people's vocabularies.[32] Science, he wrote, presents boundless opportunities. For example, scientists knew that Earth received enough solar radiation every day to power all of the steamships on the ocean and the machinery on land for thousands of years. "The only difficulty," he wrote, was "how to concentrate and utilize this wasted energy." And waste was not acceptable in the super-efficient Progressive Era, whereby any idle resource was being "wasted": standing timber, river water that was not diverted for irrigation but flowed to the ocean, deer that weren't being killed and eaten . . . all "wasted."[33] (In the mid-twentieth century, weather control advocates used the same kinds of language to justify their efforts: untapped clouds floating overhead held "wasted water.") While not claiming that it would never be possible to make rain, Newcomb argued that people had to employ adequate means to make that happen. Scientists should be able to determine in advance if rainmaking proposals could work or not. If not, why try?[34]

Newcomb also pointed out that no one understood how water particles floating in the atmosphere joined into raindrops. Smoke particles might aid the process, in which case it might be smoke rising from battlefields that triggered rain, not the concussions. "If this is the case," Newcomb wrote, "then by burning gunpowder and dynamite we are acting like Charles Lamb's Chinamen who practiced the burning of their houses for several centuries before finding that there was a cheaper way of securing the coveted delicacy of roast pig."[35]

What dismayed Newcomb was the US government's lack of effort to obtain "expert scientific evidence" before proceeding with technological solutions to problems. The question of expertise, particularly scientific expertise and its use by the state, were typical of the Progressive Era milieu. As science and technology took on increased importance, the state came to rely on men of science and engineering to provide expert testimony before making science-related decisions.[36] In this case that had not happened, much to Newcomb's chagrin. Conceding that scientific investigators tended to be "quiet, unimpressive men" who did not excite the public and who were "wholly wanting in the art of interesting the public in their work," he concluded that not even the most distinguished scientists of the eighteenth and nineteenth centuries—Lavoisier, Galvani, Ohm, or Maxwell—could have pried even a tiny appropriation out of Congress to make their great discoveries. After all, they had not dealt with projects as captivating as "attacking the rains in their aerial stronghold with dynamite bombs."[37] Newcomb, one of the few late nineteenth-century American scientists with an international reputation, remained unimpressed by the sound and light show taking place in Texas.

Equally unimpressed were physicists attending an American Association for the Advancement of Science (AAAS) meeting in Washington, DC, while Dyrenforth was exploding gas-filled balloons and rackarock. Several told the *Washington Post* that they favored experimentation, but thought Dyrenforth's experiments would have been more scientifically meaningful had he fired off explosives when conditions did *not* favor rainfall. And they reiterated Newcomb's point: the dust from the blasts may be the trigger, not the concussion. The physicists questioned whether explosives could *stop* rain, and whether the benefits of using them outweighed the costs. But others were still intrigued by Powers's "rain follows battles" ideas: perhaps moisture rose with smoke or evaporating blood from the battlefield. A water supply expert opined that if Dyrenforth and his team had produced rain, then they had "accomplished wonders." He pointed out that the resulting water supply would be worth millions of dollars in states such as Texas and Kansas, but experiments should be performed in the spring when any precipitation following explosions would be a "case of *propter hoc*, instead of *post hoc*."[38]

The press was less magnanimous than the physicists. Science "can admit neither big medicine nor prayer among her 'motors,'" wrote Walter J. Grace in the *North American Review*, but that was not true for the agencies supporting the Texas experiments. The researchers might think that rain would be shocked out of the clouds, much like nuts showering down from a shaken tree, but the better scientific explanation was that the explosion expanded the air, which then cooled to the condensation point and the cloud appeared.[39] And from the editorial board of the *Washington Post*: if it rains after Dyrenforth's experiment, how will anyone know if it was caused by the explosions or not? [This argument resurfaced in the mid-twentieth century.] Although Dyrenforth never claimed that he caused the rain, he did claim to have "encouraged" it. Reviewing the theory on the cooling of expanding air, the editorialist continued, "This theory may or may not be scientifically accurate, but it is always wisest to ascertain that the facts are beyond controversy before attempting to settle their philosophy." As for the oft-repeated statement that rain falls "immediately and invariably" after large explosions, the *Post* suggested that people credited its veracity without having attempted to verify it.[40] The *Boston Sunday Herald* editorialists poked fun at the Dyrenforth expedition and the officials who were trying to disavow it: "it never had no father." The highly regarded Department of Agriculture claimed it had been given the responsibility for the expedition, and being perfectly helpless, had passed it on to General Dyrenforth. The editorial concluded: "Gen. Dyrenforth himself, when you finally front him, is potent with thunder lightning. You don't know why, but you are sure he is charged with something, and you remain in his presence constantly on the outlook for him to _____ off."[41]

Lack of press and scientific enthusiasm aside, a month later Dyrenforth and his team conducted experiments in El Paso, Texas, which were witnessed by public, military, and US Geological Survey officials. In over ten hours they triggered seven thousand cubic feet of oxohydrogen gas, one thousand pounds of dynamite and rackarock, and one hundred twenty-one-pound bombs. The result: no rain at the explosion site. However, some rain did fall some twenty to thirty miles to its south, east, and northeast, about which the El Paso newspapers wrote favorably.[42]

Scientists and the press looked askance at such precipitation following an entire day of explosions, but the upbeat Senator Farwell told a *New York World* correspondent, "I think the experiments have now demonstrated the soundness of my theory. For twenty years I have had no doubt rain could be produced in that way, and quite expected the experiments to be successful." Looking into the future, Farwell anticipated that the Agriculture Department would be requesting annual appropriations for rainmaking—along with the usual ones for hog inspections—of between $500,000 to $1 million ($12 to $24 million in 2015, not small change). Just as Agriculture had teams of inspectors tracking crop conditions, moisture inspectors would advise the secretary on where and when rain was needed, and he would respond by sending in "men and appliances and make the rain." *Scientific American*'s editors, however, did not see rain in the future; they saw "the extraction of money from the public treasury" as the practical result.[43]

Division of Forestry Chief E. B. Fernow was not impressed either. He had declined to spend the rainmaking appropriation, calling the experiments a "waste of public money."[44] Fernow argued that the theories behind rainmaking were still "incomplete and unsatisfactory," and it was impossible to know how any given action would affect it. To say that the explosions had not affected precipitation in Texas would be "presumptuous," but there was "no reasonable ground" to say they were effective. The Texas experiments had proved nothing. It would make more sense to run laboratory experiments first before trying these techniques in the atmosphere.[45]

Famed Harvard geographer William Morris Davis also weighed in on the theoretical problems presented by both natural and artificial rainfall, and the need for both in the US West. Land in the western plains would be worth a lot more money if water were available, and the same was true for the far West. People living in the East, Davis argued, were not much concerned about water. But once they moved west, they might have sufficient water one year and too little the next. Therefore, they often eagerly embraced any plausible theory that might lead to rainfall, even to the point of asking Congress to support experimental tests. "While it is certainly not creditable to congressional action to undertake experiments upon the artificial production of rain in our present knowledge of meteorology, it is, perhaps," Davis wrote, "not surprising in view of the arguments that affect congressional action that several thousand dollars should have been appropriated for such a purpose." Unfortunately, Davis argued, those experiments were undertaken based on Edward Powers's ill-conceived writings when there are men just "as sincere as he is . . . and much better informed upon subjects bearing on meteorology." And current science did not justify pursuing this line of inquiry.[46]

So while Farwell and other congressmen thought it was worth a few thousand dollars to "test" Powers's rainmaking ideas in Texas, and then to develop even more expensive long-term programs to increase western land values, members of the scientific community were shaking their heads in dismay at the lack of connection between scientific knowledge and the experimenters. Fifty years later, more science would be present, but once again, congressional involvement muddied the water more than it cleared the smoke—literally and figuratively—surrounding the rainmaking fray.

STATE INVOLVEMENT: HELP OR HINDRANCE?

With spatial and temporal distance from Dyrenforth's explosives-palooza, Smithsonian meteorologist Curtis considered these rainmaking attempts in a social and political context for *Engineering Magazine*. He noted that even if explosives did trigger rain showers—and there was no definitive support for this contention—the experiments had not proven that it was possible to produce measureable rain in sufficient quantities to increase the economic value of semi-arid regions. Dyrenforth and his team had expended several thousand dollars, had not modified the climate, and were not close to doing so. "But the mere waste of nine thousand dollars," Curtis wrote," would be of small consequence if the effect of the enterprise could be confined to the coffers of the treasury." What worried him was the experiments' effect on the average citizen's thoughts about the possibilities of "fruitful meteorological investigation." State-funded experiments took on more importance than those funded privately. An individual who disregarded scientific counsel and conducted experiments to test impractical schemes received little to no public attention in the days before the Internet. But when the government did it, people assumed it was backed by rational thought. After all, the average person did not have the scientific expertise required to weigh a project's merits. In this case, Curtis pointed out, a single member of Congress managed to push through the appropriation without the endorsement of USDA scientists or of any "reputable

[US] physicist." Nevertheless, the average citizen assumed that the USDA had initiated this investigation and therefore the responsibility for its outcome had been placed on the government, its scientists, and on science in general.[47]

While it was wonderful, Curtis argued, that people were confident of the government's scientific work, that confidence should not be abused. Echoing Newcomb's thoughts on expertise, Congress should "hear the opinions of competent specialists before engaging in doubtful scientific projects." Similarly, scientists should not recommend that government agencies undertake experiments and projects that would not withstand the "closest scrutiny." When Curtis had been traveling in Texas for the Dyrenforth experiments, he had been struck by the complete faith people had in government, science, and the "honesty and sincerity of the Government's agents" in carrying out the project. They were convinced of the possibility to produce rain that would allow their farms and ranches to flourish and improve their way of life, even as meteorologists attempted to make clear that no rational theory or accumulated data supported such an outcome. To think that loud noises would bring rain, Curtis wrote, was to revert to an earlier, less civilized time when folk beliefs held sway over rational thought.[48]

The public's perception of the rainmaking project had not been aided by the first sensational newspaper accounts. Late nineteenth century "tweets" hyped the results: a test blast one day caused rainfall the next. ["*The Hon. C. B. Farwell, Chicago*: Preliminary. Fired some explosives yesterday afternoon. Raining hard today."] The breathless reports that followed the experimental firings sped down telegraph lines throughout the United States and abroad. But as more reasoned reports emerged, the press became less accepting and more skeptical, leading to accounts ridiculing Dyrenforth and his team. Unfortunately for the USDA and government meteorologists, the greatest number of people had read the initial dispatches and not the full accounts disputing them. Curtis bemoaned that most people thought the experiments had been successful and explosive-induced noise could produce rain on demand. "So error which will require years of teaching to eradicate," he wrote, "has been sown broadcast in a single summer, and the rainmaking myth is added to the numerous errors about the weather which already prevail." These promising accounts of rainmaking success had opened the door to "charlatans and sharpers" who were establishing "artificial-rain companies" and contracting with farmers to produce rain in semi-arid regions. Meanwhile, the plodding scientific investigations that had the greatest hope of advancing the understanding of the atmosphere and its processes had become even less likely to attract the average citizen's interest.[49]

Curtis was writing in 1892; published today, his words would still ring true with the meteorological community. When it comes to atmospheric sciences and weather control, public and scientific views have not changed much. Despite scientific journal articles pointing out the fallacies of explosives-as-rain-creator experiments, the public was not interested in pessimistic statements. Or to paraphrase the *Peanuts* comic strip's Lucy van Pelt, they didn't want any "downs, just ups, ups, and ups!" Three years after the initial experiments, the USDA continued to receive inquiries about artificial rainfall. When the volume threatened to overwhelm his staff, the Agriculture secretary created a "stereotyped letter form" advising interested citizens that the experiments had not been successful and there was no justification for farmers or anyone else to pursue them. "In this determination, judgment, and opinion," Secretary Morton wrote, "I am supported by the scientists and other alleged experts in Meteorology, connected with the US Weather Bureau." Explosive-induced concussions were not a "commercially successful" way to induce precipitation.[50] And yet interest in rainmaking continued unabated. And a number of "rainmakers" stood ready to make it rain.

RAINMAKING . . . DREAMS AND SCHEMES

Dyrenforth had had state patronage, but other rainmakers were happy to sell their unregulated services directly to local citizens. What they shared was secrecy of process and materials, much like alchemists of old. For the most part, expensive concussive blasts were not the tool of choice for commercial rainmakers who plied their trade around the country. Among them were Wyoming's Frank Alberson and Ohio's Frank Melbourne, both of whom claimed to trigger rain from buildings with holes in their roofs that allowed unknown agents to float into the sky.[51]

While Alberson, Melbourne, and others moved around the country "making rain," others continued to argue that it was the presence of increasing numbers of humans in the US West that was changing the climate. In the nation's "Great American Desert," people had cultivated the land and thereby changed the soil conditions so that rainfall soaked in instead of running off, leading to more evaporation, cloud creation, and precipitation. Planted trees produced the same result.[52]

In 1912, cereal entrepreneur C. W. Post wrote about his rainmaking ideas and efforts in the popular *Harper's Weekly*. His rainmaking interests derived from his inability to get irrigation lines to his withering crops in Texas. "It was under the stress of these conditions," Post wrote, "that I resolved to carry the war into the country of Jupiter Pluvius and bombard him until he surrendered enough rain to save the crops." "Battling" the clouds, he claimed that only the "fusillade directed at the heavens" could have produced the rain that fell. And it was economically sound as well. The rains had extended over four hundred thousand acres at a cost of one-fourth cent per acre—much less than the cost of installing and maintaining an irrigation system.[53] This very same argument would be made in the mid-twentieth century as weather control promoters championed it as a less expensive alternative to massive irrigation projects in the US West.

Post, however, was not interested in science, just in results. He was convinced that shocking rain out of the sky would "revolutionize" farming in water-short areas. "It will," Post wrote, "make the southwestern section of the United States the choicest farming region on the face of the earth."[54] Well . . . not exactly. But the idea showed amazing persistence throughout the twentieth century.

And in the face of that persistence, publications like *Scientific American* continued to push back, opining after one of Post's tests that "the rain-making hallucination is, apparently, one of the incurable forms of mental disease."[55] And it periodically tried to debunk the "rain follows battles" idea that would not fade away: "Once in the early Stone Age somebody remarked to somebody else that rain frequently occurred after battles . . . the evolution of the idea was probably complete long before the Age of Bronze." The editorialist was disappointed that he had never been able to launch another suitable cause-and-effect idea for the production of rain: "A big sneeze is often followed by rain." At least a sneeze increases local humidity, costs nothing, and is easy to find. As Simon Newcomb had pointed out, clapping one's hands in a steam-filled room did not trigger a shower.[56]

Despite the scientific press's clear contempt for rainmaking, inventors and newspapers did not stop touting the effectiveness, or potential effectiveness, of various rainmaking schemes. In the US West and Canada, probably the most famous—or infamous, depending on one's point of view—person trying to modify the weather was Charles Hatfield, "the rainmaker." That he used "secret" chemicals did not detract from his operations, since many noted that industries often used "secret processes" and they were not expected to share them. He was just successful enough at observing the weather and acting at appropriate times that his fame spread to England, which was suffering from a water shortage, not so much because of a lack of rain but because of a lack of rain in the right places. Hatfield could not only make rain, he could clear fog. And he told the *Illustrated London News* that he wanted to "dispose" of his system to the British and US governments simultaneously so that they could take over his rainmaking system and handle their own water problems. Not mentioned: the price of "disposal."[57]

The contention that producing additional rainfall was just a matter of "assisting" nature would reappear in the mid-twentieth century when rainmakers sold their proprietary methods as "enhancing" rain, not "making" it. In so doing, they seemed to be providing, as was Hatfield, a cloak of respectability in light of scientific skepticism. But most of the rainmakers were dealing with farmers and ranchers. By the 1920s, the US military would be supporting rainmaking for its own purposes. The state was back in the game.

THE ROARING TWENTIES

In contrast to the short-lived state-funded project to trigger rainfall on parched Texas grasslands in the early 1890s, the 1920s witnessed the first multiyear military-assisted effort to control the weather. Instead of focusing on producing rain by explosions, this new project initially focused on spreading electrified sand from aircraft to "bust" clouds and/or fog in support of military aviation, which had grown tremendously since World War I. But whether for creating rain where it was desired, or eliminating clouds or fog where they weren't, meteorologists remained skeptical and dismissive of their efforts.

Organized in 1919, the American Meteorological Society (AMS) was a new scientific organization with an immediate challenge: how best to inform the public about the fraudulent nature of rainmaking. In its official journal, the *Bulletin of the American Meteorological Society* (*BAMS*), A. H. Palmer held that preventing the "wasteful expenditure of public money, or the foolish investment of funds by otherwise intelligent people in projects which are fraudulent," was a worthy objective. Adopting the flamboyant Charles Hatfield as a poster boy for rainmaking charlatans, Palmer disputed claims that efforts to assist naturally occurring precipitation yielded more rain. The AMS owed the public a scientific explanation. "While it would be too much to attempt to inform the masses concerning all forms of scientific quackery and charlatanism," Palmer wrote, "it ought at least be possible to reach city officials and more intelligent farmers. . . . Only the ignorant masses through some pseudo-scientific trickery allow themselves to be deceived."[58]

A month later, *BAMS* was back with updates on Hatfield's latest ventures. In May 1920, the Ephrata (Washington) Commercial Club had hired Hatfield to produce rain in Grant County, east of the Cascade Range in Central Washington. The payment scheme was as follows: no compensation for the first inch, $3,000 for the second, and $3,000 for the third. Earlier, the farmers had approached the Grant County extension agent, who had declined to endorse this undertaking. In turn, he had contacted a "reliable source" at the University of California, Berkeley, extension office to find out more about Hatfield. By mid-July, Hatfield had produced no rain in the eastern Washington town, prompting a rival to write from Indiana, offering to produce the rain if he would not be penalized for failure.[59]

And there had been other suggestions, not only for producing rain but also for modifying the climate. The *Boston Sunday Herald* reported that "Science [*sic!*] at last has a definite plan for bringing the nice, warm Gulf Stream to our New England coast. Result—New England will have a climate like that of the Carolinas." That was too much for Boston-based AMS president Charles F. Brooks, whose letter to the editor explained, "ocean currents and climates are ruled by forces too great for control by men." He chastised the *Sunday Herald's* editors for publishing the opinions of "engineers" on topics that were better addressed by oceanographers and climatologists. Could they not check with men of suitable authority—for example, those included in the latest edition of *American Men of Science*—and print an opinion that was scientifically sound?[60]

But while the AMS meteorologists were convinced that rainmaking, cloud busting, and fog dispersal were not on the horizon, the US Army Signal Corps decided to take a chance on dissipating fog. In summer 1921, Signal Corps officials began discussing the use of electrically charged sand to clear runways and airfields of fog. The idea: determine the charge of fog droplets, spray them with oppositely charged fine grain sand, and watch the fog disappear so that planes could land safely. Seeking a professional opinion, Signal Corps major William R. Blair asked University of Minnesota physicist W. F. G. Swann to comment.[61] After providing scientific and technological insights, Swann added in a postscript that it would be rather easy to get charged particles: hang a long tube from the aircraft, slide the particles down the tube, and they would be charged by induction due to the atmospheric potential gradient.[62]

It was not quite that easy. Businessman L. Francis Warren—an interesting character who falsely claimed to hold a doctorate and be associated with Harvard College—had been collaborating with Cornell physical chemistry professor Wilder D. Bancroft to develop a viable technique for charging fine particles. In turn, Warren had contacted Harvard physicist E. L. Chaffee, who was investigating the properties of several types of particulate matter, for example, more finely divided sand, marble dust, clay, and cement dust. Chaffee advised Warren that the finer particles could hold a greater electrical charge.

Including a couple of news clippings on rainmaking in his letter, he noted that there seemed to be "plenty of money for the one who can with a certainty produce rain."[63] And for Warren, making money was the primary motivation for his weather-changing venture.

The Army Air Service—barely airborne on its much-reduced postwar budget—was extremely interested in the possibility of eliminating fog from landing strips, which were typically large grassy fields. In the early 1920s, good landing zones for aircraft, military or not, were rare. In the pre-instrument flying days, pilots needed to see the ground to land safely. Fog or dense, low clouds could force them to seek an alternate landing site. However, if they were running low on fuel, they might not be able to make it.[64] The solution? Get rid of the fog or clouds. L. Francis Warren was delighted to be of service.

At the Army Air Service's invitation, Warren traveled to McCook Field in Dayton, Ohio, to try out his proposed technique, which used a static machine to charge and disperse up to two hundred pounds of sand every three minutes. Chaffee's job was to identify the most appropriate type of sand and to create the static equipment.[65] The earliest success occurred in November 1921, when an airplane dispersing charged sand into clouds triggered a snow flurry.[66] By mid-May 1922, the researchers were struggling with material and equipment problems.[67] By late June, using military aircraft and, in this case, a Navy pilot, they were back in the air attempting to break-up clouds with charged sand. (See fig. 1.2.) At six thousand feet, the pilot "attacked" a cumulus cloud approximately three miles long, a half-mile wide, and five hundred feet deep. Flying 110 mph about fifty feet above the cloud, the pilot released sand over the full length of the cloud, continuing past its outer edge to see the effects: the cloud "immediately divided into two parts," and was mostly destroyed. He attacked the cloud three more times, ultimately causing the cloudbank to disintegrate. The pilot thought he could have destroyed the entire cloud had he been carrying enough charged sand. Observing no precipitation, he concluded the dry air surrounding the cloud had absorbed its moisture.[68] The military pilots were undoubtedly impressed; thick clouds were a flight safety challenge.

In February 1923, after sixteen months of steady experimental work, Bancroft and Warren made a public statement that was picked up by the *Times of London* and the *New York Times*.[69] The page one story in the *New York Times* reported that Bancroft and Warren were planning to move their operation from Dayton, Ohio, to storm clouds over the Atlantic Ocean in an effort to trigger precipitation. They chose the over-water location so as not to cause flooding. The *Times* reporter stressed the importance of being careful when presenting

FIGURE 1.2. Making Rain with Electrified Sand. *Science and Invention*, v. 10 n. 12, April 1923. Wilder Bancroft papers, #14-8-135. Division of Rare and Manuscript Collections, Cornell University Library.

such claims to the public "because rain making is a subject in which quacks have reveled for more than a century." One needed solid evidence to be credible. In this case, Army officers were providing supporting statements.[70] Who could be more credible?

The military pilots were not only credible; they used very vivid language to describe their experiences. One officer described how a cloud had melted "as breath into the wind." A Navy pilot maintained that he was "not particularly gullible," and indeed had been very skeptical of the whole cloud-busting idea. In fact, he had been convinced that it was another situation where "a fool and his money are soon parted." After the demonstration? He was a believer.

Warren forthrightly credited the military services for their assistance; they were providing the aircraft. He told the *Times* that the pilots were eager to help out, but they were "handicapped by insufficient appropriations." The air service was "starved" and "the decisive part which it will probably play in future wars is overlooked."[71] The lack of funding was true. Rapid demobilization after World War I had left the Army and the Navy with too little money, and too few qualified people, especially those associated with the growing field of aeronautics.

The possible use of this cloud-busting effort to garner additional appropriations for the air service was not lost on the Department of War. The Office of the Chief of Air Force wrote to Chaffee, encouraging him to travel to Washington, DC, and let people know about the experiments underway in Ohio. Publicity? The War Department had a "very complete machinery . . . to accomplish publicity on any subject," and newspapers from around the country were "clamoring" for details about the "cloud dissipation scheme." The timing was especially good; the congressional term was ending, and newspaper agencies were filing their reports. Farmers were interested in rainmaking too, and the Army Air Service thought it a "good idea to get this before congressional minds," which it was trying to focus on the importance of aviation and its military role. Indeed, its publicity shop would make Chaffee "the greatest recognized scientist in the world," and guarantee him a medal of old "Rain in the Face."[72]

Electrified sand might be used to clear dirty air over cities, remove fog from harbors, and "brighten the day . . . around dawn and sunset," but the emphasis was on military applications. The Navy wanted the capability to clear fog banks that could hide enemies or clear a path through foggy seas. The downside? What if "enterprising governments abroad" pursued these techniques and the United States neglected them?[73] How might that affect our military readiness? While this might have seemed like a logical argument for the Army and Navy flyboys in Dayton, it would have been a tougher sell to military and government leaders in the early twenties. Hadn't they just concluded a war to end all wars?

Consequently, Warren, emphasizing economics, proposed his method as a way to prevent droughts and "make deserts bloom." The *Times* gushed, "of course the discovery is destined to change geography and history, to remake the maps and later the future of the human race," although neither Bancroft nor Warren had claimed success for anything other than clearing fog and clouds.[74] But not everyone was convinced, as the cartoon in figure 1.3 illustrates.

By late summer 1923, the electrified sand project was in disarray. Money and tempers were short, vendors were clamoring for payment, and Warren was blithely urging project members to "say your prayers regularly and tighten

MAKE YOUR OWN RAIN.
Magnetized Sand Does It.

—Powers in the New York *Evening Journal.*

FIGURE 1.3. A cartoonist's take on the electrified sand method of rainmaking. *Literary Digest*, March 17, 1923, 25.

your belts at least one hole a week." He was more concerned about readying a demonstration run for flamboyant airpower proponent General Billy Mitchell and a number of foreign officers. Because Warren had no contract with the US government—he billed the Army Air Service for expenses—he urged his colleagues to be patient and "work in harmony."[75] Chaffee, under tremendous pressure to produce the equipment, suffered an emotional collapse and remained incommunicado. His associate had to single-handedly produce the nozzles used to spray the sand.[76]

In early 1924, Chaffee was back on the job, and Bancroft and Warren continued their experiments, shifting some of them from Ohio to the Aberdeen Proving Ground in Maryland as they dealt with equipment and financial difficulties. Since Warren had not been completely forthcoming with the Army Air Service about the cost of Chaffee's special-made equipment, when the money started running out in fall 1924, Chaffee was left unpaid for his time and unreimbursed for his equipment.[77] Indeed, Harvard had charged the Army less than its standard rate because it was a new project connected with the federal government. To keep the project going, Harvard had absorbed the costs—and it wanted to be reimbursed.[78]

Nevertheless, the Army demonstrated the cloud-busting technique by "shooting down" clouds over Bolling Field (at the time sharing a grass landing strip with Naval Air Station Anacostia), within sight of the US Capitol. Two Army aviators spread fine silica into a cloudbank, which dissipated in their wake. They thought the technique might be useful for clearing an approach path to a runway. The Associated Press reported that President Coolidge was interested in the experiments and hoped that they might be continued. Warren argued that the demonstration was just an extension of earlier experiments, and that with larger planes (the demonstration run used aircraft that could only release thirty pounds of sand per minute) that could spray 1,100 pounds of sand per minute he could modify frontal systems, not just small fog banks. (Note the size of the aircraft in fig. 1.4.) However, "commercial rainmaking was within the grasp of man," and one of nature's cheapest commodities—silica—could be used to make rain for $3/ton. The small cost of silica, Warren said, would be the only cost in addition to maintaining the equipment and operating the aircraft.[79] Considering that maintenance and operating costs would far exceed any amount paid for silica—which was probably not clear to the average reader—Warren seemed to be disguising the true cost of the project.

The *New York Times* revisited the project just a month later, discussing the possible civilian and military uses of the electrified-sand technique for conquering "the shortcomings of nature." The experiments had become "so

FIGURE 1.4. Aeronautical pioneer Col. Billy Mitchell at Bolling Field, site of the electrified sand experiments, in 1925. The US Capitol is visible in the background, just above the "5" on the fuselage. National Photo Company Collection, Library of Congress.

important . . . to the whole science of aeronautics and *weather control*" that the federal government had been financing them, with the Army Air Service pilots being the first to dissipate clouds before the general public. Although Army planes could carry only about two hundred pounds of sand, dirigibles could carry tons of sand, thus making possible widespread casting of electrified sand over larger cloud systems. In the meantime the author envisaged a day when a "flotilla of pilots goes aloft in the teeth of a storm" and "man first defies the clouded brow of Jove and challenges his thunderbolts in mid-air." As the Army and Navy became "extensively equipped with dirigibles," these advance parties of electrified-sand-bearing pilots would be able to clear the way for a safe landing during low visibility conditions.[80] With the federal government backing the experimental work, the obvious applications to war fighting and the possibility of turning large arid sections of the country into lush gardens were looking entirely plausible. Why have the tool and not use it to benefit the state?

The Warren-Bancroft-Chaffee experiments eventually ended. Army funding evaporated and a patent dispute over equipment between Chaffee and Warren dragged on into the 1930s without a happy resolution for anyone. The Dust Bowl years dried up people's faith in rainmaking, atmospheric moisture, and the nation's crops. Not even the country's more prominent rainmakers could make rain out of nothing at all.[81]

But while the *New York Times* reporters may have asked too few tough questions and been blinded by the promise of plentiful water in arid lands, some of the most prominent meteorologists of the day were not dazzled by the rainmakers—and they pushed back by writing books for the general public.

TAKING ON THE RAINMAKERS

Briton Sir Napier Shaw, one of the most distinguished meteorologists of the early twentieth century, took on the rainmakers in his 1923 book *The Air and Its Ways*, writing: "The control of weather is engineering. It is no more meteorology than the building of the Channel tunnel is geology."[82] So already in 1923, we see the split between meteorologists and engineers (as opposed to the charlatans) over whether it was even possible to control the weather. During World War I, Shaw had directed the British Meteorological Office and noticed weather control schemes burbling up to the highest levels of government. Those making the pitches argued that "the enemy had learned to produce rain at pleasure" and the government had a duty to "go them one better" by using some proffered apparatus. Of special note were the methods that purported to dissipate cumulus clouds by shooting them from below, when it was quite obvious to anyone fighting on the front that there was no correlation between battles and rainfall. Shaw wrote that gunfire produced two types of effects: physical (due to explosions and thermal expansion) and chemical (due to burning material). The thermal effects paled in comparison with the sun's influence, and the chemical effects were insignificant compared to those of pollutants being belched skyward by coal-fired industries.[83] As Blue Hill Observatory's Alexander McAdie suggested, it might be better to place bets "on the weather rather than on the rainmaker."[84]

McAdie published his own book, *Making the Weather*, a few months later, arguing that meteorologists did not yet know how clouds were made, and therefore ideas for modifying them were premature.[85] Scientists needed to know a lot more about cloud physics before they could possibly determine the cause and effect of rainmaking techniques.

The Weather Bureau's W. J. Humphreys entered the fray in 1926 with his book *Rain Making and Other Weather Vagaries*. Placing "rain control" schemes into three categories (magical, religious, and scientific), he provided many examples of magical and religious methods before discussing so-called scientific means. Humphreys reminded readers that they already modified their "personal weather" through clothing choices, and by using heating and cooling devices. Small-scale modification in gardens and orchards already existed; people covered plants on cold nights, used smudge pots to keep plants from freezing, and irrigated the land. Dissecting every weather modification idea since Plutarch, Humphreys pointed out that they were nonsensical, scientifically bogus, too expensive, or humanly impossible.[86]

Naturally occurring precipitation would have to suffice, and meteorologists needed to do plenty of theoretical work before they could even think about controlling it. While the 1920s had seen increased state interest in weather control with funding to match, meteorologists in particular had pushed back against the idea that an underfunded scientific discipline that had a difficult time getting the raw data needed for predicting atmospheric conditions would soon be able to change weather on demand. In the 1930s, meteorologists—influenced by the Dust Bowl—embraced cloud physics and actively sought to develop the theoretical underpinnings of precipitation processes. But would advancing theory be enough to advance control? And would meteorologists be interested in doing so even if they could? As the British fog clearer Sir Oliver Lodge told an audience at London's Institute of Physics, "There are meteorologists who know far more about the atmosphere than I do. They will, I expect, be conservative in their estimate. It may be that physicists rush in where meteorologists fear to tread."[87] As the century moved forward, physicists would not be the only ones rushing in to control the weather.

THE THIRTIES: RESEARCH INTO PRECIPITATION PROCESSES

Scientists had a fairly good grasp of cloud formation processes by the end of the nineteenth century—moist air parcels rose, for example, when they were heated by the sun and moved upward relative to surrounding air, or were carried up a mountain slope, expanded because air pressure decreased with elevation, cooled because of the expansion, and formed clouds as the water vapor cooled to the condensation point. But the water droplets only had 4×10^{-9}-inch diameters, and the smallest precipitation droplets had diameters of at least 0.04 inch. How did they form larger droplets? They didn't, unless they found their way to a hygroscopic (water-attractive) particle, for example, a particle of sea salt, dust, or sulfuric acid droplets formed from combustion. But even if they found such a particle, assuming that they were floating around together, it could take weeks to form a droplet large enough to fall. That could not be correct. Under suitable conditions, people could watch clouds form and produce rain in just a few hours.

Dutch meteorologist August W. Veraart, who conducted artificial precipitation experiments and published his results in 1930, held that the technique could be used to reduce hail storms, relieve drought conditions, bring more sunshine to cities and thus improve health, keep floods under control, extinguish forest fires, disperse fog at airports, clear clouds away from astronomical observatories, produce snow for ski resorts, and create clouds that would reduce radiational cooling at night, thus preventing crop-killing frosts. He also thought it would be more convenient if rain only fell at night, so people could go about their day without dodging raindrops. Almost two decades later, US Weather Bureau meteorologists deemed his work "not too thorough." His ideas were roundly criticized because no one had confirmed his precipitation totals and the experiments lacked sufficient controls.[88]

Two hypotheses posited processes whereby a raindrop could form faster. One, labeled drop capture, assumed that larger drops, falling faster than smaller ones, picked up water molecules on their way down through the cloud. The faster they fell, the more drops they picked up, the larger they grew, and then, when they could no longer be held aloft by upward-moving air, they dropped to earth. The second, which relied on vapor transfer, had been proposed in 1911 by German geophysicist Alfred Wegener during his research on hoar frost. Wegener suggested that if ice crystals were present in a cloud packed with minute, supercooled water droplets (i.e., the liquid drops would have a temperature below the freezing point), the latter would be drawn to the crystals until they were, once again, heavy enough to fall out. Wegener's frost research had attracted scientific attention, but the possible extension of his work to precipitation processes had not.[89]

A more serious examination of precipitation processes resulted in the groundbreaking work published by Swedish meteorologist Tor Bergeron. Fascinated by clouds since his childhood, he read Wegener's work and applied it to the precipitation problem. In February 1922, just before leaving Sweden to work with Vilhelm and Jacob Bjerknes at the Geophysical Institute in Bergen,

Norway, Bergeron had vacationed at a mountain resort. Hiking on a trail cut through the forest, he noticed that if the air temperature were below freezing, then the supercooled stratus clouds shrouding the hillside did not fill the path and he was hiking in clear air. However, in above-freezing temperatures, the clouds lowered all the way to the ground and became fog. As Bergeron thought about this puzzle, he became convinced that ice remaining on tree limbs in subfreezing temperatures pulled moisture from the cloud, leaving clear air below. But when ice on the limbs melted at higher temperatures, the cloud reached the ground.

Although much of Bergeron's time in Bergen was taken up by making weather forecasts, he collected additional data about how ice crystals affected cloud development. His 1927 doctoral dissertation—published the next year in the Norwegian journal *Geofysiske Publikasjoner* (Geophysics Publications)—gave a detailed account of his ideas on ice crystals, but received limited attention in the United States and England.[90]

In 1933, Bergeron represented Norway at the International Union of Geodesy and Geophysics (IUGG) meeting in Lisbon, Portugal, and presented a detailed paper on his ice crystal theory. He argued that in a supercooled cloud containing a few ice crystals, the latter would attract the supercooled droplets, growing larger and larger until they fell out. Bergeron concluded that all raindrops started as snowflakes (even in the summer) and fell to earth as snow if the air temperature were cold, and melted and became raindrops if it were warm. His ideas became a major topic of discussion at scientific meetings and his paper was frequently cited in academic literature. While meteorologists working in middle and high latitudes concurred in his conclusion, those working in the tropics vehemently disagreed that ice crystals were a major factor in rain production.[91]

In the late 1930s, German meteorologist Walter Findeisen provided additional measurements and calculations that helped to refine Bergeron's theory, and the ice crystal process of rain formation became known as the Bergeron-Findeisen process.[92] It remained the most widely accepted precipitation mechanism until World War II when meteorologists were confronted with reports from military aviators flying in tropical areas which forced them to look for different mechanisms in regions with significantly higher air temperatures aloft. Further research revealed that while the Bergeron-Findeisen process was plausible in middle latitudes during the winter, summer temperatures at higher altitudes were not cold enough for the process to work. The resulting collision-coalescence process posited that approximately one in 1 million droplets was larger than others in its vicinity, which allowed it to fall faster than the surrounding drops and to pick up the smaller drops as it crashed into them. When the drop was large enough, it fell out.[93]

While Bergeron and Findeisen were conducting research in Europe, Massachusetts Institute of Technology (MIT) meteorologist H. G. Houghton and his research team were conducting tests on fog dissipation for the Navy's Bureau of Aeronautics. Instead of the electrified sand of the 1920s, Houghton was planning to spray specially formulated calcium chloride powder provided by Dow Chemical on fog at a Boston-area airfield and then collect the resulting droplets on a "fog-water collector"; we would call it a screen. They intended to suspend the powder in solution held in large tanks and then to tow the tanks and accompanying spraying devices around the airfield with a truck. In at least one test carried out in 1934, visibility increased from five hundred feet to over two thousand feet in three minutes—a significant improvement for a fog-bound pilot. Houghton's quarterly report filed in August 1936 had been sufficiently positive that Navy admiral Ernest King (the former chief of the Bureau of Aeronautics) and Army general Oscar Westover (chief of the Air Corps) thought the project should continue.[94] Therefore, military interest in weather modification continued, at least as it was connected to fog dissipation and/or dispersal.

Research on cloud physics and precipitation mechanisms continued throughout the twentieth century as weather radar became more sophisticated and provided detailed information on cloud formation and behavior, while additional research focused on air pollutants and topographic influences. But in the 1940s, US researchers began taking a close look at ways of introducing nuclei to nonprecipitating clouds to produce rain or snow where desired. So did scientists in the Soviet Union, having earlier established the Turkmenistan Institute of Rainfall for carrying out rainmaking experiments by introducing chemicals—burning them on the ground or spraying from aircraft—to "wring adequate rain from them, virtually producing rain from a cloudless sky."[95]

Weather control may have been elusive, but the explosive-laden plans of the nineteenth century were giving way to science-based weather control in the mid-twentieth century. Both US and Soviet state interests provided a substantial boost to these efforts.

CHAPTER 2

Weather in an Icebox: Scientific Weather Control

We may yet have rain or sunshine by pressing radio buttons.

DAVID SARNOFF, Chairman, Radio Corporation of America

September 30, 1946, New York City. Speaking at a glittering testimonial dinner honoring his forty years of service to radio, Radio Corporation of America (RCA) president David Sarnoff told the one thousand distinguished attendees that weather control was a scientific possibility. "For example," Sarnoff declared, "man may learn how to deflect air movements with consequent changes in weather and he may discover how to neutralize a storm or detour it from its course." Eventually people might be able to choose their weather much as they chose a radio station. When that day arrived, he continued, "we shall need a World Weather Bureau in which global forecasting and control will be vested."[1]

Sarnoff was giving voice to the postwar euphoria that embraced a science-and-technology-can-fix-anything approach to all of America's problems, natural or man-made.[2] But he was not the only person discussing weather control. RCA physicist Vladimir Zworykin was promoting his idea of using an electronic digital computer to forecast the weather and ultimately control it. And when his colleague, internationally known mathematician John von Neumann, submitted a proposal to the US Navy to develop numerical methods of weather prediction using his still-on-the-drawing-board electronic digital computer in spring 1946, weather control was the anticipated outcome.[3]

While Zworykin and von Neumann had pie-in-the-sky visions of how weather control might work as their "virtual weather" became "real-world weather," Nobel Prize–winning chemist Irving Langmuir and his team at the General Electric Research Laboratory in Schenectady, New York, were actively seeking ways to make it happen. As laboratory researchers, they were convinced that what they controlled in the lab could be controlled in the atmosphere. Meteorologists—for whom the atmosphere was a massive, complex, uncontrollable natural space containing too many variables to track—vehemently disagreed.[4] And for postwar-America's military and civilian leaders willing to exploit any advantage in the nation's increasingly adversarial relationship with the Soviet Union, the possibility of ensuring national security by controlling the weather was too tantalizing to pass up.

If we consider these three groups—weather control innovators and promoters, Weather Bureau meteorologists, and federal government civilian and military leaders—as occupying the nodes of a triangle, the partnerships and tensions between the resulting pairs provide a window into the development of weather control techniques and policies starting in the immediate postwar years. As each group jockeyed for position in early 1947, federal funds and military materiel flowed to weather control research as science-and-technology-can-fix-anything ideals infused the sciences, engineering . . . and the state.[5]

SMOKE AND ICE: THE GENERAL ELECTRIC LAB

The GE lab, site of technological innovation and scientific development since its 1900 founding by Thomas Alva Edison and several others, had been Irving Langmuir's scientific home since 1909. A man of tremendous intellect and wide-ranging curiosity, Langmuir had researched numerous phenomena including lighting, reactions at high temperatures and low pressures, thermal effects in gases, chemical forces, and electrical discharges in gases. Starting in the late 1930s and continuing until his death in 1957, he also turned his attention to phenomena outside the laboratory. During World War II, he created cloaking smokes and the equipment to produce them, and he improved gas masks for the Army. In 1943, Langmuir and his assistant, Vincent Schaefer, undertook a military contract focused on the buildup of rime and clear ice on aircraft, which entailed studying cloud droplet size and crystal formation.[6] Langmuir, noticing that moisture-laden clouds did not always precipitate, hypothesized that they lacked "ice nuclei."[7]

While they studied what caused some clouds to precipitate while others did not, Schaefer obtained a GE home freezer in which he planned to introduce a supercooled cloud and artificially created ice crystals. To clearly see crystals a few microns in diameter, he lined the freezer compartment with black velvet and directed a strong beam of light into it. The temperature was -23°C (-9°F) at the bottom of the freezer, and Schaefer created a "cloud" by exhaling into it. He then tested chemicals one after another, sprinkling a few particles into the freezer in hopes of producing ice crystals from the cloud. The few he saw were always near the bottom of the freezer. Frustrated by his lack of success, in July 1946, Schaefer rapidly cooled the freezer by adding a block of dry ice (with a surface temperature of -78.5°C or -109.3°F). He was stunned by a flurry of ice crystals that filled the freezer box before fluttering to the bottom. Eagerly pursuing this line of experimentation, Schaefer dropped smaller and smaller bits of dry ice into the box, generating a miniature snowstorm each time. Indeed, any substance chilled to -40°C (-40°F) produced the same result: ice crystals formed about one millimeter apart for a total of 100 million ice nuclei in the freezer.[8]

Langmuir moved Schaefer's empirical research onto theoretical ground by studying the growth rate of nuclei produced when dry ice pellets were dropped through supercooled clouds. He concluded that the limiting factor in "seeding" (adding dry ice pellets to) clouds was not the number of nuclei, but their distribution rate. According to Langmuir, when supercooled liquid water droplets evaporated, the amount remaining as gaseous water vapor was considerably less than the amount that condensed onto ice nuclei. The condensation process added latent heat to the cloud, triggering upward vertical motion and turbulence. Although the turbulence eventually retarded the upward motion, it brought in air from outside the cloud and increased air movement within. The lateral and vertical spreading of seeds throughout a stratus (flat) cloud would, Langmuir thought, lead to its complete nucleation—and precipitation—in about thirty minutes.[9]

To test Langmuir's hypothesis, Schaefer and another GE scientist climbed into a single-engine Fairchild airplane and took off from the Schenectady airport in search of promising clouds. The morning's stratus clouds were dissipating, so they flew until they found stratus of temperature near -20°C (-4°F) at an altitude of about 4,300 meters (fourteen thousand feet). Thirty miles away from the airstrip, Schaefer scattered three pounds of dry ice shavings out of the cockpit window along a three-mile-long line. Atop the airport's control tower platform, Langmuir scanned the clouds with his binoculars and watched as snow fell from beneath the seeded cloud. Three minutes later, the flat stratus cloud began growing cauliflower-like bulges extending upward almost five hundred feet before disappearing into a "veil of snow." Within five minutes of seeding, the entire cloud had been transformed into snow that fell about two thousand feet toward the ground before evaporating into the dry autumn air.[10] Schaefer made several more test runs, all of which successfully turned large clouds into ice crystals. When he seeded a cumulus cloud, it was completely transformed into ice crystals.[11]

Based on their observations, Langmuir concluded that a plane flying two hundred miles per hour could seed one thousand square miles of clouds in an hour. Since the GE scientists could dissipate a single cloud layer in less than thirty minutes by seeding from above a cloud for landings and from below for takeoffs, a plane carrying a seeding apparatus could create a hole large enough to fly through, thus preventing aircraft icing. In addition to this military application, Langmuir thought that seeding subfreezing clouds on the windward side of mountains would augment rain- and snowfall, producing more runoff for summer irrigation. Another possible application: seeding the tops of cumulonimbus clouds ("thunder clouds") to make them less destructive, thereby reducing the annual $15 million in hail damage in the US West.[12]

While Schaefer experimented with dry ice, another member of the research team, atmospheric scientist Bernard Vonnegut, had been identifying additional seeding agents that mimicked ice nuclei structure. Silver iodide was the winner; dissolving it in a strong potassium iodide solution and diluting it with acetone yielded a solution into which small charcoal briquettes could be dipped. Once dried and burned, the briquettes released about 100 trillion silver iodide nuclei (or seeds) per second. Using silver iodide ground generators (fig. 2.1) in windward mountain areas, upslope winds carried the seeds into clouds, inducing precipitation at negligible cost.[13] Silver iodide seeds had another advantage over dry ice: they did not evaporate, remaining airborne much longer.[14]

Langmuir, Schaefer, and Vonnegut thought they had viable techniques for weather modification, but many questions still needed to be addressed. What characteristics did "foreign" nuclei have, and how effective were they? What types of nuclei produced the best results? How susceptible were different cloud types to seeding? What was the distribution of droplet sizes in clouds? What were cloud conditions before rain or snow began to fall? What happened if cloud tops were seeded?[15] To answer these and other more practical questions, the team needed to conduct basic research, which required a patron with deep pockets, aircraft, and the resources to protect General Electric from liability if an experiment went awry. The US military services were pleased to help.

THE RESEARCH AND DEVELOPMENT BOARD TAKES ON WEATHER CONTROL

Like many other scientists who had supported the war effort, Langmuir had close connections to the military services, and less than two weeks after his initial tests the dry ice-based cloud-seeding technique attracted the attention

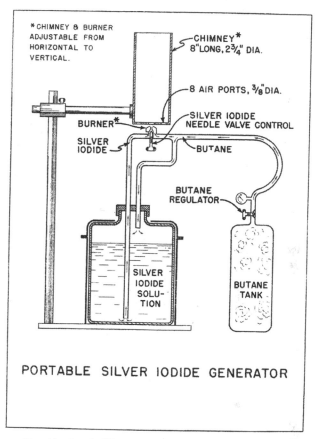

FIGURE 2.1. Portable silver iodide generator. From C. R. Holmes and William Hume II, "Final Research Report to the State of New Mexico Economic Development Commission Water Resources Development," New Mexico Institute of Mining and Technology, September 1951.

of the military's Research and Development Board (RDB) and its chairman, federal science policy guru Vannevar Bush. The RDB had played a vital role in the military's wartime use of science, and Bush later became the principal science adviser to Secretary of Defense James Forrestal. Bush's task was to bring civilian scientists and military officers together to develop policy related to the research and development (R&D) of new weapons,[16] and he had more than academic interest in Langmuir's snowmaking experiments. If a few pounds of dry ice could dump tons of snow on an enemy, then this method had serious military applications. (As fig. 2.2 indicates, the military services were interested in a wide variety of earth science applications in support of their activities.) Bush asked the RDB's Committee on Geophysical Science to

(R)   5. The field of the geophysical sciences is broad and the military
applications of these sciences are numerous. The unsolved problems in
such general areas as oceanography, meteorology, geology are of a fundamental nature. The panels of the Committee have formulated the boundaries
of the unknown in their respective fields and have rated the importance
of the solution of unsolved problems from a scientific and military viewpoint. This was done in order to help in the formulation of a master plan
of research and development in the geophysical sciences. This task, for a
fundamental field of science, is complicated by the multitudinous military
applications. The following table shows a few significant examples of the
general relations of scientific fields to military objectives.

a.	Carrography & Geodesy - - -	Missile ranging and guidance problems; military mapping; terrain models.
b.	Geology - - - - - - - - - -	Strategic minerals; terrain intelligence.
c.	Hydrology - - - - - - - - -	Water supplies; floods; military construction on ice and permafrost.
d.	Meteorology - - - - - - - -	Weather forecasting for air operations; weather control in land and air operations.
e.	Upper Atmosphere- - - - - -	Guided missile design; long range communications.
f.	Atmospheric Electricity - -	Protection of aircraft radio communications.
g.	Oceanography - - - - - - -	Underseas warfare.
h.	Seismology - - - - - - -	Shock protection of surface and subsurface installations; hurricane detection.
i.	Soil Mechanics - - - - - -	Vehicle trafficability.
j.	Terrestrial Magnetism and Electricity - - - - -	Mine and submarine detection; guidance system for missiles; degaussing

FIGURE 2.2. The Research and Development Board was very interested in the connections between geophysical sciences and military applications. This list of geophysical applications includes weather control under meteorology. From RDB Committee on Geophysical Sciences Annual Report, July 1, 1947–June 30, 1948, June 15, 1948 [SECRET], Box 227/4, RG 330, RDB Entry E341, NARA II.

explore weather control, assess its military implications, and recommend the best way to supplement basic research in atmospheric processes with broader studies related to military applications.[17] The investigation's leader: Carl-Gustav Rossby, chairman of the Panel on Meteorology, who had spearheaded the training of thousands of American meteorologists during World War II. Of equal importance: the entrepreneurial Rossby was probably the most prominent meteorologist in the world at the time, and he commanded tremendous professional respect, making him an ideal leader for a group examining a very controversial topic.[18]

The committee's interim report addressed the long period of controversial scientific development that had preceded Schaefer's successful experiment, and the need for additional experimentation to determine the implications and possible exploitation of weather control, especially for military purposes. The ability to determine in advance the location and timing of precipitation was important to military planners. Other possible uses included temporarily dissipating solid cloud layers, thereby permitting pilots to locate targets or use airfields for landings and takeoffs, and reducing flight hazards, such as icing and static, that could lead to lightning strikes or instances of St. Elmo's Fire, an electrical discharge that may occur when the magnitude of the environment's electric field is high, which can destroy aircraft radar domes. Despite the uncertainty, the secretary of war considered seeding's possible military uses to be so sensitive that he maintained complete control over related press releases, and Army and Navy representatives classified practical applications of cloud seeding.[19]

Meteorology panel members concluded that cloud seeding showed economic and military promise, and they recommended funding R&D related to it, keeping the basic science unclassified since much of it was readily available and openly publishing research results to speed up development. Many Western scholars thought that the German meteorologist Walter Findeisen, who had dropped out of sight during the war, had fallen into Soviet hands and thus it would be fruitless to keep precipitation processes classified—echoing the same concerns that had been expressed over nuclear fusion in the United States and the cancer-fighting KR preparation in the USSR.[20] Panel members also discussed possible legislation to control cloud modification, perhaps patterned after irrigation law, with the federal departments most likely to be affected by weather control—War, Navy, Commerce, Agriculture, and Interior—forming a joint oversight commission.[21]

While weather control's scientific considerations and operational applications remained under review, the military services "peddled" its technical possibilities to operational personnel, seeking potential tactical uses.[22] They responded enthusiastically. Artificial rain could hinder enemy forces, precipitating snow away from urban areas could reduce snow removal expenses, and inducing precipitation over agricultural areas could ensure sufficient soil moisture for increased crop yields. Future uses with the greatest long-term implications included diverting precipitation and exhausting clouds' water content.[23] The Army advocated clearing clouds to aid photoreconnaissance, foiling the use of cloud cover by enemy defenses, and opening "holes" near airborne forces' drop zones. Weather control could reduce enemy troop morale by inflicting excessive, periodic inclement weather on staging and resting areas, destroy the enemy's will to continue conflicts by damaging its harvests, and trigger snow to expose camouflaged emplacements and reveal signs of enemy activity on supply routes.[24]

These potential military uses were classified, but the science behind them was not. Increasingly, it found its way into the popular press, and oft-exaggerated media coverage sparked civilian queries for assistance to break droughts, snuff out forest fires, and bust hurricanes. Since the military services were funding weather control for military, not civilian, applications, the RDB referred civilian requests for weather intervention to the US Weather Bureau. Funding for civilian applications had to come from other sources, while the military budget for weather control grew along with its possible operational uses.[25]

After one year of investigating weather control, the RDB had no answers on its efficacy for the Joint Chiefs of Staff; however, it argued that potential uses fully justified intensive R&D efforts.[26] The many scientific, technical, legal, and political difficulties needed to be addressed by a variety of agencies during the research process. In the meantime, Langmuir's research project had already secured military funding. It was called Project Cirrus.

THE US ARMY SIGNAL CORPS CONTRACTS—PROJECT CIRRUS BEGINS

The US Army Signal Corps signed its inaugural contract for fog- and cloud-dispersion research with General Electric in February 1947. Initially dubbed the "snow project" and later renamed Project Cirrus, the contracts continued through September 1952 at a cost of almost $800,000 (approximately $6.5 million in 2015 dollars). In their joint statement, the Signal Corps and General Electric announced that their research might lead to the "manipulation of gigantic natural forces for the benefit of mankind everywhere." GE's vice president for research, C. Guy Suits, cautioned that widespread weather control was probably far in the future, but that under the right conditions it might be possible to change clouds and induce precipitation wherever and whenever desired.[27] The "wherever" and "whenever" attracted the most military interest, and military officials made the trip to see GE's laboratory experiment for themselves (figs. 2.3 and 2.4).

Not everyone was thrilled. Climatologist Helmut Landsberg, deputy executive director of the Air Weather Service's R&D branch, requested a copy of the contract, but its arrival failed to shed light on the classified *reasons* behind the contract. A Signal Corps representative reminded Landsberg that it would not have awarded the contract had it not expected to obtain usable scientific information. If the Army could find a way to trigger precipitation processes, its forces could precipitate clouds before they could interfere with military operations, dissipate them to remove cover from enemy staging areas or targets, or precipitate clouds on enemy troops, hampering ground operations. "Making weather" was nothing more than taking advantage of nature for military purposes.[28]

Despite the Army's practical needs, Project Cirrus was not solely focused on applications. Its scientific purpose was to increase understanding of the physics and chemistry behind the formation of hydrometeors (rain, snow, hail, drizzle, etc.) and thereby improve weather forecasts for military operations. Researchers would examine cloud microstructure, determining cloud particles' water content, their size-based distribution, and clouds' vertical growth rate. They would attempt cloud modification with dry ice and other experimental nuclei to produce rain or snow, using smoke bombs during initial trials to differentiate between treated and untreated stratus clouds.[29]

The contract included Langmuir and Schaefer's services, but to protect GE from liability for perceived damage from induced precipitation, military personnel would conduct the field tests. Neither the Army nor GE would release results without permission from the War Department.[30] The lack of indemnity for contractors conducting state-sponsored weather control experiments in the open atmosphere would contribute to efforts to pass weather control legislation starting in 1951 (see chapter 3).

By mid-July 1947, Langmuir was analyzing flight results, establishing experimental procedures, and planning methods and techniques for future programs. He was also developing a mathematical theory to address how air movement within clouds affected the growth of cloud particles, water droplets, and ice crystals. Schaefer was conducting laboratory experiments on basic processes, assisting Langmuir, and keeping track of cloud studies worldwide. Vonnegut was developing ground and airborne silver iodide generators,

FIGURE 2.3. Members of the Army Signal Corps observe the "ice box" experiment on March 13, 1947, at the GE Laboratory in Schenectady, NY. Courtesy of the Museum of Innovation and Science, Schenectady, NY.

FIGURE 2.4. The iconic photograph of (clockwise from the top) Irving Langmuir, Bernard Vonnegut, and Vincent Schaefer with the icebox. Courtesy of the Museum of Innovation and Science, Schenectady, NY.

determining meteorological conditions that were most favorable to ground generator use, and assisting the Navy with a fog generator. Others concentrated on instrumentation and gathering data.[31]

The GE team members took a three-pronged approach, conducting laboratory, photographic, and field studies. In the laboratory, they studied supercooled clouds and how seeding affected the production of ice crystals and their growth rate. They also developed field kits to detect ice nuclei, and they examined the properties of clouds under high electric fields. The photographic team took time-lapse photographs to provide evidence of changes in seeded clouds. Field studies included detecting supersaturation (i.e., relative humidity greater than 100 percent) by creating cirrus clouds in clear skies by launching balloons carrying dry ice in open mesh bags and tracking their paths (fig. 2.5). Sublimation trails (much like jet contrails) left behind by the floating dry ice bags provided evidence of supersaturation.[32]

Based on early results, Langmuir concluded that their seeding techniques held huge possibilities for modifying weather: altering cloudiness over the northern United States and providing sunshine to the upper Midwest during winter, decreasing cloud cover, preventing ice and freezing rainstorms, and reducing aircraft icing. By reducing cloud cover, they could lower albedo (i.e., the amount of reflectivity), increase the amount of heat absorbed from sunlight, and thus raise the average air temperature of some regions. Langmuir proposed trying these techniques in sparsely populated Alaska or northern Canada before trying them out in the lower-48 states.[33] There is no indication in the archival records of what Alaskans or Canadians thought of this plan.

Team members continued testing various dry ice granule sizes (fig. 2.6) depending on cloud type and the effectiveness of silver iodide smoke under different atmospheric conditions, and running experiments to determine how seeding modified clouds, how seeds propagated through clouds, and how long the effects lasted. In mid-November, Langmuir announced another method for producing rain: dispensing water droplets into actively growing cumulus clouds to start a "chain reaction" within the cloud. The droplets, which were large compared to the cloud droplets, would fall through the cloud, picking up moisture along the way until they got so big that they broke into smaller droplets, which would be carried aloft within the cloud to start the process again. Ultimately, the entire cloud would produce heavy rain. Not every cloud was susceptible to the charms of water droplets. The ideal cloud had to have five miles per hour of upward vertical motion, fully grown cloud water droplets, a high water content, and several thousand feet of thickness—the kinds of clouds found in the US Northeast in summer, and year-round in the southern

FIGURE 2.5. Weather balloons carrying mesh bags filled with dry ice during artificial nucleation experiments in New Mexico. From C. R. Holmes and William Hume II, "Final Research Report to the State of New Mexico Economic Development Commission Water Resources Development," New Mexico Institute of Mining and Technology, September 1951.

F I G U R E 2 . 6 . The Project Cirrus team loading dry ice into boxes before a cloud-seeding run. Courtesy of the Museum of Innovation and Science, Schenectady, NY.

Pacific Coast states and the tropics.[34] Experiments conducted by the Pineapple Research Institute in Honolulu, Hawaii, also contributed to Langmuir's "chain reaction" theory, and his hypotheses were tested later in Puerto Rico during the Thunderstorm Project.[35] However, it was much more difficult to figure out what was happening during field experiments than during lab experiments—just as meteorologists had always argued.

LEGISLATION: DIRECTION AND CONTROL

The military-funded Project Cirrus was seeking basic scientific knowledge and practical weather control applications when Representative Sid Simpson (R-Illinois) introduced a bill to appropriate $500,000 to investigate turning clouds into rain. The proposed legislation, H.R. 4582, directed the chief of the Weather Bureau to conduct experiments concerning methods of controlling

rainfall and perfecting methods of triggering and controlling rainfall for speci-
fied areas. Apparently unaware of ongoing military involvement, Simpson also
suggested that the experiments might have military significance. A skeptical
RDB member noted on his file copy: "This ought to buy a lot of rattles and
snake skins."[36] A half-million dollars was a lot of money, but Simpson's grasp
of science was tenuous at best if he thought it would enable the perfection of
weather control.

During the March 1948 committee hearings, General Electric's representative discussed his company's involvement in cloud seeding and argued that
bringing weather control to fruition would cost a lot more than $500,000.
However, lawsuits, not money, were GE's major concern. The bill's language
allowed the Department of Defense (DoD) to indemnify contractors, but GE
wanted protection from multimillion-dollar lawsuits. Their larger legal questions included: To whom does cloud-borne water belong? Does any individual have the right to precipitate water? If water may be precipitated, where
and when may it be done? Who is liable for any resulting damage?[37] Directing
weather control research had more far-reaching implications than just producing rain on demand. And based on GE's questions, they all concerned water.

The Weather Bureau bristled at Congress's attempt to dictate its research
program, but federal agriculture-related agencies were positively disposed toward weather control research. Secretary of Agriculture Clinton P. Anderson,
eight months shy of being elected New Mexico's junior senator, supported the
bill because of rain's "vital importance" for crops in semi-arid and drought-
affected humid regions. So did the Agriculture Research Administration
(ARA) and the US Forest Service (USFS).[38] The ARA argued that rain falling
in the right place at the right time and in the right amount would be terrific.
But since induced rain might fall on those who did not want it, the federal
government should carry out the experimental work.[39] The USFS desired increased humidity reducing fire threats and aiding fire suppression. Dissipating
thunderstorms and preventing lightning would be a plus. It recommended
experimenting in remote areas of the Pacific Northwest: very few people lived
there and the forests would benefit.[40] The Weather Bureau realized it could not
talk its way out of weather control research, so it took the money and applied
it to basic cloud physics.

THE CLOUD PHYSICS PROJECT

In May 1948, the Weather Bureau, Air Force, and the National Advisory Committee for Aeronautics (NACA) launched the Cloud Physics Project, directed

and staffed by USWB meteorologists.[41] The project's mission was to produce or suppress precipitation on demand and to increase visibility for aircraft. To differentiate between artificially induced and natural weather effects, project members installed instruments designed to continuously record air pressure, temperature, humidity, wind velocity, and precipitation at fifty-five ground stations covering 160 square miles near Wilmington, Ohio. They also had access to two upper-air stations and two "high power ground radars": the former launched instrument packages (radiosondes) that measured temperature, pressure, and relative humidity as they were carried aloft by large balloons, while the latter provided a continuous photographic record of the radar returns. The team monitored all aircraft traffic in the area, recorded all precipitation, and used airplanes for both seeding and photography.[42]

That project director Ross Gunn, a meteorologist, and his colleagues were dubious about the ability of seeding to trigger massive rainfall is an understatement. Just two weeks into the project they had their evidence: artificially induced rain never fell unless natural precipitation was falling within thirty miles of the seeded area. *There were no exceptions.* Seeding had not produced economically significant rainfall nor had it initiated self-propagating storms.[43] Over the next few months, during which time they also had access to airborne radar that could estimate precipitation rates, they found no additional data that changed their minds. The researchers could convert nonprecipitating clouds filled with supercooled water droplets into snowing clouds in winter, but so what? Under no conditions could they "extract more water from clouds or the air than contained therein." However, Gunn and his team still wanted to examine the effects of orographic lifting (forced lifting of air by mountains) and spot-check the effects of seeding on clouds in other parts of the country so that their critics could not argue that the results were valid only in Ohio.[44]

Six months into the project, in late November 1948, the Air Force unexpectedly dropped out of the project and into a buzz saw of snarky press criticism.[45] A *New York Times* editorial wondered what had happened to the ideas of alleviating droughts and steering hurricanes away from populated areas, and the newspaper cheekily forecasted a "dry spell" in rainmaking. GE's hometown newspaper accused the Air Force of making a very old mistake: thinking that what couldn't be done in the present couldn't be done in the future. Its riposte: "Wanna Bet?"[46]

The joint USWB/USAF report, overshadowed by the untimely press release, reiterated Gunn's earlier conclusions: seeding clouds was not economically viable and it did not cause storms. Only large masses of moist air carried in by winds that induced large-scale cooling produced significant amounts of rain or snow.[47] Weather control's two main desired applications—significant precipitation on demand for domestic use and tactical weather control for the military services—were not likely to result from the research underway.

The Australians, working on their own cloud physics project, begged to differ. A *New York Times* front page story, which likely sent Weather Bureau chief Francis W. Reichelderfer right over the edge, reported that the Australian research team not only thought rainmaking had real value, but that it could identify which clouds were most amenable to seeding.[48] Therefore, the Australians could guarantee significant results from cloud seeding. The ever-skeptical Reichelderfer, realizing that press stories about scientific topics could often be unreliable, wrote to the director of Australia's Meteorology Bureau, asking for its unfiltered scientific results. He advised the Australian meteorologists that some of the Cloud Physics Project's tests might have been labeled "successful" had it lacked radar evidence showing naturally occurring rain just outside the seeding area.[49] Reichelderfer emphasized the importance of looking at the larger meteorological picture before judging cloud seeding's effectiveness, lest they conclude that their actions had produced rainfall when they had not.

With this mostly political flap behind them, the Cloud Physics Project's second phase focused on the practical limits and economic importance of cloud modification processes, specifically the effectiveness of dry ice, lead oxide, and potassium iodide seeding of cumuliform clouds. Again, the overarching problem was separating naturally occurring from artificially induced precipitation, and once again, artificially induced rain was more likely in the presence of naturally precipitating clouds. Seeded cumulus clouds, however, tended to dissipate rather than precipitate—not a promising outcome for alleviating drought conditions by seeding "fair weather cumulus" dotting brilliant blue skies as the ground dried beneath them.[50] Hence, the Weather Bureau investigators held firm to their original conclusion: seeding-induced precipitation was not economically important.

While the Commerce Department–based Weather Bureau dealt with economics, the Air Force researchers focused on their primary meteorologically based desire: all-weather flying. Could they modify clouds to eliminate problems created by low cloud ceilings, poor visibility, severe icing, and turbulence? Could they dissipate thunderstorms, which could ruin a pilot's day and aircraft? In their preliminary report (labeled RESTRICTED), they reached the same discouraging conclusions as their Weather Bureau counterparts, but did uncover two possible benefits for aviation: seeding individual cumulus clouds that were growing larger often led to rapid dissipation, which would prevent thunderstorm formation, and seeding might also dissipate stratus clouds, but to a lesser degree.[51] Of the nineteen experimental runs at the Wilmington site, not one significantly changed cloud behavior.[52] The Air Force's overall assessment of weather control techniques: not overwhelmingly positive.

Air Force and Weather Bureau meteorologists uncovered basic science underlying cloud physics, but little evidence to bolster hopes of mitigating weather-related aviation hazards or making drier portions of the country bloom without huge irrigation projects. Nevertheless, they moved their experiments to a variety of spots around the country: Brookhaven, New York, to run tests on supercooled fog when it proved too cold to conduct seeding experiments in Fairbanks, Alaska, and later to the West Coast to seed clouds heading east from the Pacific Ocean.[53]

While the Cloud Physics Project team had been working in Wilmington, Ohio, Langmuir and the Project Cirrus team had set up shop in New Mexico's high desert. His conclusion would be diametrically opposed to that of Reichelderfer and the Cloud Physics Project team: not only was seeding economically important; it was possible to control weather all across the continent with minimal seeding. Those conclusions would spark a very public tiff with Reichelderfer in early 1949. But before we engage with that turf battle, let's check in with Project Cirrus.

LANGMUIR PROMOTES WEATHER CONTROL—PROJECT CIRRUS IN NEW MEXICO

Frustrated that the Cloud Physics Project was holding fast to its conclusion that cloud seeding did not produce economically significant amounts of rain, Langmuir doubled down as he busily promoted Project Cirrus and its results to anyone who would listen. A crucial part of his message: he had never encountered a supercooled cloud that could not be converted to ice crystals by adding dry ice particles. With that outcome assured, Langmuir was convinced that it would be possible to eliminate conditions that produced dangerous aircraft icing by converting all supercooled cloud droplets into harmless ice crystals. He was so sure of success, he wanted to scale-up his experimental area to roughly five hundred square miles and conduct a large-scale test just off the coast of Alaska.[54]

Langmuir was also conducting seeding experiments on white, puffy cumulus clouds, particularly *cumulus congestus*, which are typically characterized by sharp outlines and extensive vertical growth and are commonly known as "towering cumulus." Frequently associated with torrential downpours, given

FIGURE 2.7. Mobile silver iodide generator mounted on an automobile for seeding experiments in New Mexico. From C. R. Holmes and William Hume II, "Final Research Report to the State of New Mexico Economic Development Commission Water Resources Development," New Mexico Institute of Mining and Technology, September 1951.

favorable conditions cumulus congestus can grow into cumulonimbus (i.e., "thunder clouds"), instantly recognizable due to their fibrous, ice-crystal-laden anvil-shaped tops. Conducting field experiments in New Mexico (some of which used mobile seeding generators like the one in fig. 2.7), Langmuir was especially interested in the effects of seeding on towering cumulus clouds that had not yet become cumulonimbus. He claimed that seeding towering cumulus tops made "things happen very fast" and produced unusual phenomena. While he was driving under a seeded cloud in New Mexico, the rain had come down in "torrents," forcing him to pull off the road. The accompanying violent wind—probably a downdraft associated with the dissipating stage of the thunderstorm—knocked over trees. Then came the hail, "small pieces" of one-half-inch diameter, which may have seemed unusual to Langmuir, but were not rare in New Mexico. Langmuir, wondering why the seeded clouds had precipitated out so rapidly, surmised that a "chain reaction" had occurred in nature's storm, which led to the fragmentation of snow crystals and hence snowfalls. If true, it might be possible to induce larger amounts of precipitation by using dry ice to trigger the chain reaction in moisture-laden, naturally growing storms. Convinced of his ability to trigger large storms, Langmuir felt obligated to say that he did not want to trigger such a snowstorm over New York City and leave officials with a $5 million cleanup.[55] The Big Apple must have been relieved.

Langmuir being Langmuir, reporting on experimental results without developing a mathematical theory to explain them would have been unthinkable. He calculated that, given air moving upward at five miles per hour, enough moisture to produce at least one-tenth of an inch of rain, and individual twenty-micron diameter cloud particles, *a single drop of water* would cause an entire cumulus cloud to release its moisture. Indeed, in warmer tropical clouds, seeding with water droplets would be just as effective as seeding with dry ice.[56] Less expensive, too.

Langmuir also wanted to apply his techniques to forest fire prevention and hurricane steering. The former had been tried in New England by seeding clouds, but rain started falling before he could seed. The Navy was interested in steering hurricanes away from ships at sea, and, perhaps more importantly, from bases where many ships were docked and in danger of sinking next to piers. Langmuir expressed his eagerness to study hurricanes far off shore. His experimental plan was to fly back and forth over hurricanes and observe if seeding modified them or steered them onto different (and presumably less dangerous) paths. Langmuir thought increased knowledge about seeding's effects would make it possible to "abolish the evil effects" of hurricanes.[57]

Langmuir had big plans for weather control—so big that the language he used tied it more closely to atomic energy than to meteorology. His use of the term "chain reaction" to describe what was taking place in seeded clouds harkens back to the same language used by nuclear scientists to describe events taking place during atomic explosions. Indeed, we will see this same language incorporated into statements supporting 1950s-era state control of the weather. On the other hand, Langmuir also displayed his lack of meteorological understanding. His description of weather phenomena associated with New Mexico thunderstorms could have been given by almost any longtime resident of the Desert Southwest. And yet, Langmuir did not ask the most obvious question: Did I just experience phenomena outside recorded meteorological extremes? Instead, he immediately concluded that his seeding experiments caused the driving rain and fell back on his theoretical calculations to declare that a single drop of water could work as the trigger. And Langmuir implied that he could effectively produce weather on demand by referring to the inadvisability of dropping heavy snow on New York City. Similarly, his remarks about modifying (weakening) and steering hurricanes implied that he fully understood the nuances of hurricane structure in 1948—nuances that are still unknown in the early twenty-first century. Langmuir traded on his status as an elite scientist to proffer ideas that would have been quickly dismissed as being from the scientific fringe at best, and quackish at worst, had they come from just about anyone else. To their chagrin, Weather Bureau scientists were routinely left to justify their assertion that Langmuir's ability to control the weather was not ready for prime time.

At the August 1948 International Union of Geodesy and Geophysics (IUGG) meeting in Oslo, Francis Reichelderfer informed fellow geophysicists about the Cloud Physics Project's results. Langmuir was furious, but kept quiet. By January 1949, however, when Reichelderfer and Langmuir both presented papers at the annual American Meteorological Society meeting, their mutual enmity made its public debut. What were Langmuir's main points of contention?

The finding that seeded clouds were more likely to rain when other clouds were raining within thirty to sixty miles of the seeding site? Langmuir claimed the Weather Bureau was trying to "belittle" any "positive" results. A thirty-mile radius encompassed 2,900 square miles; a sixty-mile radius, 11,600 square miles—a lot of territory. "Surely," Langmuir wrote, "the occurrence of rain within such a large area cannot be taken as proof that any given cloud would have seeded itself at the time that the experiment was carried out."[58] Langmuir missed Reichelderfer's point. If clouds in the vicinity rained on their own, couldn't the seeded clouds have rained eventually without seeding? Were raining clouds significantly different than nonraining clouds? Were clouds in the same region markedly different from one another?

Reichelderfer acknowledged that clouds could be dissipated with dry ice, but doubted the possibility of making large-scale rain. When Langmuir protested, Reichelderfer claimed he was counteracting Langmuir's "grossly exaggerated statements" on the efficacy of rainmaking, specifically the ones related to controlling hurricanes. Langmuir was nonplussed. Contrary to Reichelderfer's claim that hurricanes were too large to influence with seeding, Langmuir argued that since larger storms contained more energy, seeding at the "proper stage of development" should produce widespread effects. "To assume that a hurricane could not be successfully modified by even a single pellet of dry ice," Langmuir wrote in his notes, "is like assuming that a very large forest fire could not be set on fire by such a small thing as a single match."[59] Had Reichelderfer ever seen this note, he likely would have rolled his eyes dismissively before replying, "No, modifying a hurricane with a single pellet of dry ice would be like extinguishing a very large fire with a single drop of water. Neither one of them would be effective."

The Weather Bureau's lack of laboratory tests was a continuing sore point for Langmuir, who was convinced that questions about artificial nucleation were going to be very difficult to settle by atmospheric experiments. He thought it was "far better to arrive at a thorough understanding of the meteorological factors based on laboratory experiments and suitably chosen sets of meteorological data." Then he could "lay out a theoretical basis" for the meteorological phenomena found in supercooled clouds.[60] For Langmuir, the laboratory beat out the atmosphere for research because he could *control* everything in the laboratory. He wanted to control atmospheric conditions as he controlled every aspect of his laboratory, and would learn how to do it in the laboratory. No wonder Weather Bureau meteorologists looked askance at Langmuir's scientific approach. Observational data gathered *in situ* were what counted—and they did not forecast the weather for a laboratory.[61]

And yet, when Langmuir made his case at the annual AMS meeting, he argued that GE and Project Cirrus had not, in general, been trying to make rain—especially due to potential legal problems—and then summarized the "fundamental facts" of seeding.[62] Considering that Langmuir had been extolling the possibilities of controlling rainfall since late 1946 and the military services were performing the Project Cirrus field experiments to shield GE, his comments appear a tad disingenuous.

Reichelderfer took a different tack, discussing weather control as one of three approaches (improving observations and forecasts being the others) to aid the aeronautics community. He argued that weather modification's mission was to "abolish" the weather, which sounds odd until we remember that on a delightfully fair day with virtually no wind there is "no weather." Reichelderfer was saying that weather control was about controlling unfavorable phenomena or "bad weather." It was inaccurate to say that artificially producing a desired weather condition was impossible, but some weather "controls" were more practical than others. Considering the massive volume of air and energy carried by storms, and the costs and consequences of large-scale weather control, it seemed unlikely that it would solve aviation's weather problems. Small-scale changes, such as dissipating fog near airports, were more likely to be practical and cost-effective.[63]

Reichelderfer likely thought he had given a realistic appraisal of weather control's possibilities. But the Department of the Interior disagreed, charging the Weather Bureau with attempting to prove the *impossibility* of weather control and then requesting funding for its own experiments as a countermeasure. Stung by Interior's perception, Reichelderfer pondered how to change it. As he gathered data from other research teams in the United States, Canada, and Australia, he was struck by two findings: they were convinced that the Cloud Physics Project had had only negative results (while everyone else had had some positive and some negative results), and their raw data differed little from the Weather Bureau's. Consequently, Reichelderfer and his team were perceived as being so hostile to weather control that they could not acknowledge successes, while in turn, Reichelderfer thought the others were overstating their successes.[64] How could they get their message out about the efficacy of weather control without appearing overtly negative?

THE MEDIA DEFENSIVE

Reichelderfer could not have led a charm offensive on behalf of the Weather Bureau's weather control results. He was not that kind of person. However, to get their version of the story out, Reichelderfer and his staff needed to carefully choose their words as they approached the press and spoke about weather control at a variety of venues.

Ross Gunn made the first attempt, writing an article about the Cloud Physics Project for *Physics Today*. Considering Reichelderfer's almost paranoid desire to hold information about weather control close, Gunn wisely (or not) routed it to him for comments. Reichelderfer's response: was Gunn writing as a disinterested observer, or as a partisan in the cloud seeding controversy? Reichelderfer argued that it was best to let the facts speak for themselves; press controversy just confused readers and wasted time. And, he told Gunn, leave your personality, your personal opinions, and references to the Weather Bureau out as well.[65]

Reichelderfer's comments might have made sense if Gunn had been writing for the popular press, but physicists were *Physics Today*'s audience. Gunn was not disinterested; he was party to the controversy. How could he just "stick to the facts"? Scientific facts rarely speak for themselves. Someone must interpret them. Should that person be, in this case, a meteorologist? Or did Reichelderfer think that a journalist could do a better job? And the bit about Gunn suppressing his personality and personal opinions? Did Reichelderfer really think that Langmuir's Nobel Prize–winning reputation, and by extension his personality, had not played a role in how government officials and the public had received his pronouncements on weather control? They were undoubtedly thinking: how could a Nobel Prize winner be wrong? Who do these Weather Bureau meteorologists think they are, disputing the word of a chemist in a white lab coat? Perhaps Reichelderfer clung to the comfortable notion that scientists were completely "objective," hermetically sealed from any factors that might interfere with the appropriate interpretation of experimental data. But his lack of media savvy was not helpful.

Shortly after reining in Gunn, Reichelderfer went on the offensive with a speaking-to-the-choir talk at Washington, DC's Cosmos Club—an invitation-only organization then composed primarily of prominent earth scientists—about cloud physics and cloud seeding research. After reiterating what atmospheric scientists knew and did not know about cloud development and precipitation mechanisms, he turned to Langmuir's seeding ideas. Using technical language and lots of facts, Reichelderfer's core message was this: Look, fellow earth scientists, these GE folks, who are *not* meteorologists, think they can use dry ice or silver iodide to tinker with the atmosphere and make it do what they want. But we meteorologists *know* that the enormous amounts of energy spread throughout a large volume of air will not be modified by something as small as silver iodide seeds, no matter how many of them are introduced into the atmosphere.[66] Except all those attending the luncheon already knew that. He still was not getting his message out to a broader audience. Perhaps more important: was he getting his message out to fellow government bureaucrats who needed some solid guidance?

EVALUATING "PROGRESS" IN WEATHER CONTROL

By mid-May 1949, Reichelderfer was concerned that Weather Bureau personnel were spending more time on operations and less on basic research since the Cloud Physics Project had moved from Ohio to the West Coast. He thought it important to stop "short of the boondoggling stage," which was fast approaching, and not spend time and money on tests for little return.[67] The pullback on the Cloud Physics Project did not mean that Reichelderfer wanted to cut ties with Project Cirrus. Indeed, his overarching desire was to keep a close eye on Project Cirrus and Langmuir.

Simultaneously, the Research and Development Board's Panel on the Atmosphere reported that clearing airfields of fog and military areas of clouds, increasing visibility in clouds, and eliminating aircraft icing hazards were all important military applications, and it predicted future cloud seeding effectiveness:

- In 5 years: 80 percent effectiveness in dispersing supercooled clouds and 50 percent in dispersing supercooled fogs; 10 percent of warm clouds (temperature above freezing) dispersed;
- In 10 years: all supercooled fogs and clouds would be dispersed; 20 percent of warm clouds dispersed;
- In 15 years: no further problems dealing with supercooled clouds; 20 percent of warm clouds dispersed.

The panel recommended more research on all phases of cloud physics to extend knowledge of cloud and fog phenomena, including worldwide cloud seeding experiments to include a variety of climate types.[68] Considering the lack of evidence, its predictions were at best overly optimistic, and at worst, based on wishful thinking.

Meanwhile the Weather Bureau's Project Cirrus consultant—mole is more accurate—was visiting Langmuir at the GE Research Laboratory. Reichelderfer had planted William Lewis in the project to monitor Langmuir's undertakings while providing sorely needed meteorological sense. Lewis concentrated his efforts on data from cumulus cloud seeding that had taken place in New Mexico in summer and early fall 1948. The main problem: the grid spacing between hourly reporting stations was larger than the convective systems, which tended to slip undetected between the grid points much as small fish slip through trawler nets. Contrary to Langmuir's contention, heavy rains were not unusual in the Santa Fe area and therefore provided no evidence for or against cloud seeding effectiveness. Writing to Langmuir, Lewis argued that the Project Cirrus team was facing a problem that was more complex than Langmuir was acknowledging and needed to use statistical analysis to determine seeding's significance. Furthermore, although most meteorological situations produce a wide variety of scientific opinions (as meteorologists know all too well), Lewis wrote that he had never been in a cumulonimbus cloud that did not have ice crystals. That clean-cut flat anvil top? It looked that way because of the distance between the observer and the cloud, not because of ice crystals, which likely formed far below 33,700 feet. The photographs in which the clouds looked diffuse did not offer proof of ice crystals. The clouds only looked diffuse because the photograph was taken with the dissipating part of the cloud against the light. There was no proof that the seeds' effects moved with the tallest cloud, nor was there reason to believe that clouds Langmuir thought were seed-induced had not occurred naturally. Furthermore, ice crystals could fall from cumulonimbus into lower-level clouds, effectively serving as natural seeds, which provided an alternative hypothesis for cloud development without artificial seeding.[69] In all probability, Langmuir did not appreciate Lewis's observations.

While Langmuir and his team were trying to analyze the seeding results from their New Mexico field experiments, with or without Lewis's advice, military advisers for the Cloud Physics Project decided that the next experiments should be conducted on warm season cumulus clouds. Florida was an ideal test site because a large variety of single-cell and isolated cumulus clouds were readily available.[70]

Because stratiform clouds did not produce precipitation, the military was only interested in seeding enough of the cloud to create holes for pilots to fly through. Therefore, project scientists decided to conduct cold season super-cooled stratus and fog seeding from a base in New England or Newfoundland to ascertain the effectiveness of seeding for military operations. For example, if seeding created a hole like the one in figure 2.8, did it stay open long enough for a pilot to use it? Weather Bureau meteorologists suggested that military units lead this set of experiments, while they would analyze the data.[71] The Air Force and Navy agreed and proceeded accordingly.[72]

Military meteorologists also agreed to cooperate with a cumulus-seeding

FIGURE 2.8. Military officials wanted seeding experiments to prove the viability of putting holes in stratus clouds to aid aviation. US Air Force Photo, NOAA Central Library.

project planned for New Mexico since Air Force resources were available. However, they concentrated on clearing fog and low clouds that were potential aviation hazards, leaving precipitation enhancement to civilians.[73]

CONTINENTAL WEATHER CONTROL?

Throughout the late 1940s, Irving Langmuir's comments generally addressed cloud clearing and "local" weather modification activities. After moving his experimental work to New Mexico, however, he became convinced that his local actions had continental-sized influence, and he shared his conclusions widely. Most people working with technologies realize they spawn both intended and unintended consequences. Similarly, most people realize that their words may produce unintended consequences, too. Langmuir, apparently, was not among them. As a result, General Electric soon found itself in a rather uncomfortable position.

In July 1949, Langmuir conducted a cloud seeding experiment in New Mexico that, he claimed, showed that silver iodide could produce rain large distances—thirty to fifty miles—away from the seeding site, and that induced rain would be "easily" detectable at distances of two hundred miles or more. He subsequently arranged for his team to seed in New Mexico's high desert from December 1949 through September 1950. Because it took the Weather Bureau three months to compile national monthly climatological data, Langmuir had to wait until April for the station data needed for his analysis. However, in January 1950, he learned of flooding in the Ohio and Wabash river valleys—the most serious in many years. He directed his GE staff to gather twenty-four-hour rainfall totals from relevant watershed observation stations. Struck by a weekly rainfall periodicity over the entire area, Langmuir concluded that seeding in New Mexico had been at least partly responsible for the flood-inducing rains, and he scaled back seeding from three days per week to two days every other week.[74] A few weeks later, flooding started to affect the Mississippi River. And yes, Langmuir took credit for enhancing precipitation in that case as well.

Weather Bureau meteorologists looking at the same data reached a different conclusion: the precipitation was due to natural causes. If, however, Langmuir were proved correct, the results would be of "extreme importance" to the military; an independent appraisal was needed. The Signal Corps's Michael Ference—a Project Cirrus adviser—recommended that the Research and Development Board appoint a team of consultants to meet with Langmuir and discuss his results.[75]

The consulting team members—George P. Wadsworth (Mathematics/ MIT), Gardner Emmons (Engineering/NYU), Hurd C. Willett (Meteorology/MIT), and Bernhard Haurwitz (Meteorology/NYU), all distinguished academics—visited Langmuir in Schenectady, but his mountain of data was too much for them to analyze in a single day, so they carried it home, intending to conduct statistical and synoptic checks of the data before returning for another visit. The team planned to spend only one more day with Langmuir, who thought they would stay for extended discussions.[76]

Following their second meeting, team members concluded that Langmuir's data did not support his contention that he had triggered the rain that had fallen in the Mississippi Valley. If Langmuir wanted them to accept that claim, then he needed to offer considerably more evidence. While their charge had not been to analyze the fundamental concepts of cloud physics, they pointed out that the current state of scientific knowledge made it impossible to determine from physical reasoning what happened inside seeded clouds. They also could not determine the horizontal and vertical transportation and distribution of seeding material through the atmosphere, and they could not track the seeds from New Mexico to the Mississippi River Valley. Observed rain was insufficient to prove or disprove Langmuir's claim that his seeding had spawned a low-pressure area that produced large amounts of rain almost one thousand miles away.[77] Based on the tenuous nature of Langmuir's argument, the evaluators fell back on a standard scientific recommendation: do more experimental work with better controls, followed by a robust statistical analysis of the data.[78]

Because the report disputed Langmuir's claims, it turned into a political hot potato and its release was postponed despite the RDB Panel on the Atmosphere's decision to make it public. An Army representative was particularly concerned because under the terms of the Signal Corps/GE contract, GE got a "cut" on any press releases and might issue a rebuttal. It would be unseemly for the Department of Defense to disagree with its own contractor.[79]

As military officials struggled with releasing the Emmons et al. report, Langmuir was in the tiny dust-and-scrub-filled high desert town of Socorro to address the New Mexico School of Mines' graduating class. In his speech, Langmuir tied "indiscriminate and promiscuous" seeding [by ranchers] in New Mexico with flooding in Manitoba, some 1,500 miles to the north. Fascinated Canadian officials contacted General Electric for more information. Within days, GE's General News Bureau asked Langmuir to "recall" those remarks and get them to the company as soon as possible.[80] Manitoba was not on any kind of storm track that passed through New Mexico, and the conditions leading to the floods, which included massive winter snows, had been months in the making, but those facts did not concern Langmuir. They did concern meteorologists, who were trying to keep up with, and tamp down, Langmuir's more meteorologically unsupportable public statements.

Failing to take heed when meteorologists told him that seeding clouds in New Mexico likely only affected weather within a few miles of the seeding site, an undeterred Langmuir published a paper in *Science*, one of the oldest and most influential journals in the American scientific community, that reiterated his idea that seeding near Albuquerque had triggered 0.35 inch of rain over four thousand square miles in a single day—100 million tons of rain. Upon recalculating the amount, he increased it eightfold to 800 million tons, all from seeding. He then claimed that by shooting a single pellet of dry ice from a pistol into a cloud, he could make it rain. (Just how that could be done and leave the dry ice pellet intact is a mystery.) Running with this idea, Langmuir leaped to proposing that it was "highly probable" that hurricanes could be modified and prevented from reaching land if they were seeded in an incipient stage.[81] The Weather Bureau's Ferguson Hall and Emmons and his colleagues lambasted the article in letters to the editor.[82] Writing separately, they dissected Langmuir's argument with such precision that it makes one wonder if Langmuir's article had gone through peer review, or if he had just been given a "pass" as a Nobel Prize winner.

Shortly before Langmuir's article appeared in *Science*, the RDB released the Emmons et al. report accompanied by the Army's dissent and a fact sheet. In an interesting development, almost immediately thereafter RDB chairman William Webster decided that for "security reasons" it was not a good idea to identify the Department of Defense with the implication that weather control was not going to be viable for several years, if then. Therefore, the report, especially any comments related to the DoD and the RDB, or to possible strategic or tactical inferences that could be "plainly" drawn from it, needed to be classified "confidential" and everyone possessing a copy had to mark and handle it accordingly.[83]

The Justice Department also obtained a copy of the report and queried the Commerce secretary about Weather Bureau activities related to certain "unconventional methods of attack." The Commerce secretary truthfully responded that except for "incidental" studies made by the bureau related to "weather control," it had not been involved with military activities.[84] However, the entire handling of this incident, from the investigation conducted at the RDB's behest by Emmons et al. to the debate over the report's release, followed by Langmuir's article in *Science*, indicates the level of confusion surrounding weather control's efficacy as well as its potential use as a state tool.

While mulling over the Emmons et al. report's implications, the Panel on the Atmosphere assembled its 1950 edition of the "technical estimate" of the status of military-related atmospheric research. The first draft acknowledged the importance of more accurate, longer-range weather forecasts, but weather control possibilities were gaining ground. Realistically, however, progress was slow, and being able to modify the microclimate of orchards was not equivalent to large-scale weather control.[85] The final, more circumspect version of the report suggested that the military *might* be able to modify local cloud cover and thereby induce precipitation, and modify natural storms in a small way.[86] The military services wanted to be in on the ground floor with weather control, but cooler heads prevailed in the creation of official documents. Members of the Panel on the Atmosphere did not want to criticize Langmuir, particularly because of the contracts between the Signal Service and GE, nor did they want to make unsupportable claims for weather control.

LANGMUIR STAKES HIS CLAIM

Despite the overt skepticism of atmospheric scientists, in 1950 Langmuir was still pursuing his theory that induced rainfall was proportional to the square of the amount of silver iodide used. He concluded that the chance of this enhanced rainfall being due to a random (natural) event was 1 million to one. He also concluded that seeding with a single pound of silver iodide each week would modify the US climate, and burning additional silver iodide–treated briquettes would produce more widespread effects.[87]

He was also nursing a tremendous grudge against meteorologists in general, and those of the Weather Bureau in particular. Langmuir would later tell *New York Times* reporter John Pfeiffer: "[Weather experts] tend to feel that anyone who hasn't specialized in meteorology isn't in the run. But all of us have to be prepared for the unexpected, for things that don't make sense according to our old ideas. We must continue to experiment with open minds. That's the scientific method."[88] Unless, of course, the data do not support the argument and one continues to make it anyway. In that case, an unsupported hypothesis should just fade away. Langmuir was completely convinced that his data supported his hypothesis, and the "orthodox weather experts" were unable to see it.

With his ideas spurned by the nation's "highest possible authorities," Langmuir decided to make his case at the National Academy of Sciences meeting on October 12, 1950, in Schenectady. When Langmuir presented, with "great vehemence," his paper connecting seeding in New Mexico with regular rainfall across the United States, attendees were "startled." He argued that the seven-day periodicity was "quite beyond the range of any fortuitous effects." But when asked if seeding had directly caused the rain, he demurred. "The ordinary concept of cause and effect," he declared, "cannot be applied in this case." Although many "experts" had said that such large-scale weather modification was impossible, Langmuir used terms like "astounding" and "magnificent" to describe his results.[89] Meteorologists were not impressed.

However, Langmuir did not attempt to publish the paper, and only the abstract appeared in *Science* on October 20, 1950.[90] Based on his own account, it appears that Langmuir was unwilling to expose his data analysis to peer review, especially if the reviewers were meteorologists. As a Nobel Prize winner, he seemed disturbed that his ideas were not immediately embraced and that a bunch of niggling meteorologists, so very low in the scientific pecking order, should be thumbing their collective noses at his obvious inducement of periodicities across the United States.

Curious about the paper, Reichelderfer asked Langmuir for a copy. In his fourteen-page response, Langmuir proposed testing his theory of periodic rainfall and having the USWB's statisticians determine if the periodicities were "so high that they practically indicate certainty." Then, Langmuir wrote, they could work together to determine their underlying causes. He thought that the possibility of inducing economically significant amounts of precipitation paled into *"insignificance"* compared to what he saw as the most important outcome: the complete revision of the underpinnings of synoptic meteorology.[91] Langmuir was not only convinced that he was correct—seeding in New Mexico had changed the weather all across the continent—but that his work was so revolutionary that it would change the course of meteorology. While he accused the Weather Bureau of being obstructionist, Langmuir apparently felt completely justified in ignoring hard data and analysis presented by people who had been working in the field for decades. This may have been the point when Langmuir's weather control efforts started to fall apart: arrogance set him up to look foolish at best when weather control failed to materialize.

As 1950 ended and news of the proposed Weather Control Act was hitting newspapers, United Press International quoted Langmuir as saying that weather control could be "as powerful a war weapon as the atom bomb."

Therefore, the federal government should do for weather control what it had done for atomic energy. Liberating the energy from thirty milligrams of silver iodide under "optimum conditions" would be equivalent to the energy released from an atomic bomb. Charging that the Weather Bureau opposed his ideas, Langmuir maintained that weather control could produce drought, torrential downpours, and deprive hydroelectric plants of fuel.[92]

Two weeks later, GE announced it would not enforce its patents related to the use of silver iodide to induce precipitation to "permit a maximum freedom of research in this field." Vice President and Director of Research C. Guy Suits acknowledged that the company research lab needed to accomplish much more developmental work before anyone would see the full benefits of its discoveries.[93] And the benefits would be many: the correct types and amounts of precipitation at the optimum time, the elimination of hazardous weather, and sunshine for those who remained under cloudy skies most winters.

And Langmuir? Earlier in the year he had resigned as a State University of New York trustee so he could pursue rainmaking. Langmuir said, "I believe that the best service I can render to the national welfare is to increase rather than decrease my activities in [rainmaking]."[94] He may have been convinced that rainmaking and national welfare—and defense—went together, but the Research and Development Board's committees were not yet firmly in Langmuir's camp. But they didn't need to be. Langmuir had already hooked a bigger fish, one who saw weather control as a state tool and could lead the charge for state funded weather control: former secretary of Agriculture and junior senator from New Mexico Clinton P. Anderson.

* II *

Coming to Grips with Weather Control (1950–1957)

"War," wrote sociologist and historian Charles Tilly, "made the modern state." And the Cold War, with its pervasive, imminent threat of war, solidified a "powerful central state" in mid-twentieth-century America that drew heavily on science, technology, and natural resources to create overwhelming military strength to counter and surpass the similar efforts of its rival, the Soviet Union. For the United States—a nation in which strong antistate ideals have long prevailed—a crisis is a necessary, although not sufficient, condition for triggering a state-building period. Economic meltdowns like the Great Depression count. Wars, hot and cold, count. *The* Cold War counts.[1] In the 1950s, Cold War national security concerns drove the United States to use all available science and technology to build an offensive and defensive arsenal to preserve "truth, justice, and the American way," and inspired planning for a new bureaucracy (Weather Control Commission) around weather control and its use as a new, relatively inexpensive, nontraceable, radiation-free weapon. Add in drought conditions around the country and unregulated attempts to mitigate them with weather control, and the smaller crises started to multiply. In short order, weather control looked like an ideal state tool to address both international and domestic problems—a tool that would require significant state patronage—and become yet another part of the scientific state.[2]

In Part I, scientific management and its concomitant efficiencies were central to Progressive Era thought, and the state increased patronage for governmental science and its in-house experts throughout the early twentieth century. But scientific management by the state really took off during World War II, when the entire country (scientists and engineers included) was mobilized for the war effort. As Brian Balogh puts it, "Exchanges between experts and the federal government evolved from what might be characterized as a coy but chaste Progressive Era courtship, to a triumphant marriage during the early years of the Cold War."[3] Add in some less-than-chaste flings in the Roaring Twenties—including attempts to use electrified sand to clear clouds for Army aviators—and science began shifting from simply a rational method for making progress in the world to a "national resource . . . and a ward of the state." The state supported its new ward, and in turn, society reaped the benefits. But state support inevitably comes with strings attached. Patrons always make the rules, and scientists in the 1950s found themselves struggling to negotiate this new reality as they worked to advise the state on science-related policy while working at the state's behest, adapting to the state's agenda. Science, then, played an integral role in the American state—which, in the 1950s, was not only a scientific state, but also a national security state.[4]

Science and technology had played outsized roles in the winning of World War II, and America's science maven, Vannevar Bush, who had headed the US Office of Scientific Research and Development, wanted to ensure that they continued to play an outsized role in the nation's future. His report *Science—the Endless Frontier* argued that scientific progress was of "vital interest to the Government," because without it, people would be less healthy, the standard of living would plunge because of insufficient job creation, and the nation would be unable to fight back against tyranny. In short, the state needed to use science for the national welfare because the nation's security depended upon it.[5]

With the end of World War II and the beginning of the Cold War, the Department of Defense (DoD) became *the* patron for science, a science that was no longer small, but big . . . very big. Big science generally brings to mind military investments in physics and engineering related to nuclear weapon development, but the earth sciences, including meteorology and oceanography, were not bit players during a time when 80 percent of the federal research and development budget was in the hands of the DoD. Government labs, major industrial labs (including General Electric), and science departments of major universities were on the receiving end of the DoD's largesse, which was directly related to the national security state.[6] Because "civilian strategists regularly overestimated the power and malignity of their Soviet adversary," as historian Bruce Kuklick has written—and those same strategists maintained that an appropriate measure of what American supremacy should look like was how it had looked in 1945, when the United States was the only major power left with its economy and infrastructure intact—requesting more money for scientific and technological endeavors that would keep the American state ahead of the Soviet state became a winning strategy for scientists and the agencies that funded them, particularly the DoD, Atomic Energy Commission, and National Science Foundation.[7]

Patronage, for science or any other field of endeavor, influences the direction that work takes. For centuries, artists produced religious works because wealthy donors commissioned them for churches. When military patrons provide funding for scientific research, then scientists' research agendas will be subtly, or not so subtly, steered toward answering questions of military concern over time, or they will find a new patron with deep pockets. In that way, the state exerts political control over science.[8] The 1950s-era weather control story is interesting because it fits into the overall story of Cold War weapons development without being about physicists: the military wanted to keep cloud modification research solely under its auspices. But considering the scientists who were trying to advance this work, it is not exactly a meteorology story either. While some eye-rolling meteorologists may have been willing to encourage cloud seeding research if it had solely involved enhancing rainfall, they drew the line at being involved in weapons development. Instead, they strongly argued that they didn't have sufficient knowledge of atmospheric processes to control the weather. Period. All stop.

Perhaps, had they not been so worried about their professional reputations, they might then have added, "Okay . . . you want to control the weather? Fine. First we need a large wad of money to tease apart basic precipitation processes. That will take a few years. Then, we'll need another large wad of money . . . oh, and aircraft for experimental platforms, special movie cameras, lots of film, plenty of people . . . to run years of tests on a variety of cloud types in different topographical areas. And that will just be to test clearing fog, punching holes in clouds, and enhancing precipitation. If you want us to prevent floods, blizzards, tornadoes, and/or hurricanes, that will cost you another gazillion dollars over a few decades. Let us know when you want us to start!"

But they didn't. Why not? Because leaders of the US Weather Bureau (USWB), which still employed the majority of US meteorologists in the 1950s, knew that they did not have sufficient bureaucratic autonomy to change the military's agenda or the minds of congressmen who had decided that a full-scale assault on weather control supervised by a Weather Control Commission was a way to strengthen national security and to overcome problems facing individual states who were (mostly unsuccessfully) trying to address intra- and interstate weather control conflicts.[9] As it turned out, the USWB's lack of public legitimacy and bureaucratic entrepreneurs during the Cold War doomed its personnel to fighting a rearguard action against congressional patrons who wanted to change the direction of the nation's weather service from one of forecasting to one of control. When senators holding checkbooks found the meteorologists to be recalcitrant in the face of their entreaties, scientists who were not meteorologists were willing to step in and accept the checks.

How could that have happened? If the American state were built on scientific principles and depended on disciplinary expertise, how could congressmen have ignored the meteorologists? Wouldn't they have been the experts in this case? Well, yes and no. Yes: they were the ones who knew the most about the atmosphere, and they were also the ones who knew what they *did not know* about the atmosphere and *said so*. What good are scientists who don't know everything? Isn't that their job? *That is what makes them experts.* However, the Nobel Prize–winning chemist Irving Langmuir—he of white coat and a leader at a major industrial laboratory (General Electric)—while he did *not* know nearly as much about the atmosphere and its behavior as the USWB meteorologists, emphatically and confidently defended his claim that he had controlled the weather over the entire continent. The senators were enthralled. *He was an expert.* As my students would say, he was *awesome*.

And because, as sociologists Harry Collins and Robert Evans write in *Rethinking Expertise*, "distance lends enchantment," knowledge appears to be more certain the farther one is from the creation of that knowledge. Langmuir was trying to pass weather control off as a certainty—and his congressional audience had no way to check the validity of the argument. Thus, they relied on his personal reputation. The USWB meteorologists, who could show that Langmuir was wrong, simply could not measure up. Their voices didn't count, and they were scooted over to the kids' table so the grownups (senators) could rhapsodize about a future where lousy weather was banished, and where "scientists in white coats ... [were] given license to speak with authority on almost any subject." Why was this? Because in the 1950s, scientists spoke about *all* of science with authority, even when they did not know what they were talking about—and some did it more than others, especially the weather control proponents.[10] Consequently, the congressional decisions on weather control were not informed by science, but by politics: the politics of national security in the Cold War.

As the 1950s opened, three disparate yet related groups were seeking answers to questions about weather control: federal lawmakers (and the executive departments and their subordinate agencies that sought to exploit and provide some measure of control over weather control activities); state lawmakers and agencies (seeking to referee brewing legal fights concerning intrastate and interstate weather control); and the atmospheric scientists—in other words, the subject matter experts. (As it turns out, these researchers—in government, academe, and the commercial sector—were not easily divided into pro- or con-weather control camps.)

The trajectories of these groups run in parallel throughout the decade, with occasional intermingling among them. Somewhat like a play—one where the audience sees the same event from the viewpoint of three different actors—the next three chapters explore the themes of state influence on science, examining how scientific expertise and professional associations influenced state decisions regarding science and technology vis-à-vis weather control.

CHAPTER 3

US Congress:
Controlling Weather Control

The nation that first learns to plot the paths of air masses accurately and learns to control
the time and place of precipitation will dominate the globe.

GENERAL GEORGE C. KENNY, US Air Force[1]

Cold warriors were fascinated by weather control's possibilities. By late 1950,
the theoretical underpinnings of cloud physics and the potential practical uses
of weather control had grabbed the attention of General George C. Kenny, former commander of the Strategic Air Command, who foresaw global domina-
tion by the state able to produce weather on demand. Equally enamored were
residents of the US West whose states lacked sufficient water to support the
postwar boom in population, agriculture, and industry. Although the Desert
Southwest most commonly springs to mind, cotton and tobacco growers en-
during droughts in the East, ranchers scratching out a living in the rain shadow
of Pacific mountain ranges, farmers in the semi-arid Plains, and New York
City's leaders: all were eyeing weather control to fix whatever atmospheric
problems ailed them.

While the military's weather control efforts were cloaked in secrecy, media
reports—fed by Irving Langmuir's overly positive claims of success—attracted
attention around the country. Writing in the *New York Journal-American* in
1950, reporter Louis Reid gushed, "Rainmakers will soon be taking orders—to
provide just the kind of weather you want." Langmuir, Reid reported, claimed
that science had "penetrated the secrets" of weather control. Any weather condition could be changed. Farmers and cities could get rain; skiers snow. Big
snow headed for your town? Intercept the clouds and let the snow fall "harmlessly" over a forest or an ocean. Instead of letting nature seed clouds willy-
nilly, target the seeds. Scientists could change winter cloud cover and raise the
air temperature by sending 1/1000th of an ounce of silver iodide up through
the bases of large, supercooled cumulus clouds encompassing thirty cubic

miles of air, thereby liberating more heat than an atomic bomb. Langmuir, wrote Reid, had produced 320 billion gallons of water over thirty thousand square miles for about twenty bucks.[2] While the overwhelming hype disturbed US Weather Bureau (USWB) and academic meteorologists, typical readers were led to believe that weather control did not reside in the realm of science fiction. It was already standing by to solve all of their weather problems.

STIRRING UP LEGISLATION (SEPTEMBER– DECEMBER 1950)

Suggestions for regulating rainmaking and other weather control efforts began to burble up alongside the hype, as scientists, including GE's Bernard Vonnegut, openly worried that uncontrolled seeding could hinder their ability to control experimental conditions. Senator Clinton P. Anderson (D-New Mexico), the former agriculture secretary, was receptive to their concerns, and in September 1950, Washington, DC's muckraking newspaper columnist and radio broadcaster Drew Pearson announced that Anderson would introduce a bill to establish a federal commission to regulate rainmaking.[3]

Pearson was correct: Anderson was considering legislation. Unsure how to proceed, Anderson and his aide, Edward E. Triviz, began seeking advice. The USWB told them to put regulations on the back burner until scientists understood their experimental results.[4] That was the wrong answer. They continued to cast about for guidance and landed at the Library of Congress's Legislative Reference Service (LRS), explaining that cloud seeding might benefit one area to another's detriment. Were any federal agencies vested with the power to regulate rainmaking? Was a new agency required? What existing regulatory body might be a model? Would weather control legislation be timely? Was there sufficient justification to pursue it?[5] Can we do some state building here?

A year earlier, a House of Representatives report had rejected weather control activities as being appropriate undertakings for the federal government because they would duplicate existing government and private sector work and could draw the federal government into litigation brought by states and municipalities. That same congressional session had led to the Cloud Physics Project (chapter 2). With that in mind, the LRS suggested two approaches: provide general authority for federal research and experiments or else establish extensive and complete federal control over all research and experiments. The first option would enhance collaborations among groups addressing all types of precipitation enhancement without hampering private, state, and municipal efforts and would fall under the Federal Tort Claims Act. The second would require the establishment of a new agency, perhaps a "Federal Precipitation Commission," because the understaffed and underfunded USWB lacked the administrative resources to manage a weather control program. The US Department of Agriculture (USDA), however, thought the USWB was the most logical agency to carry out weather control research. Anderson already knew that. He had made the same argument as agriculture secretary.[6]

However, at least two other constitutional clauses provided suitable legislative options: the commerce clause, since weather affected interstate traffic, flood control, and watershed development; and the war powers clause, because weather control could be used in peacetime for national defense. Individual states could not claim exclusive rights to rain and weather under the Tenth Amendment (powers reserved to the states), so that could not block the bill. Although it might make sense to postpone legislation until weather control was less controversial, the LRS suggested that Anderson could offer the use of weather control for eliminating droughts, ensuring adequate precipitation for agriculture, and preventing heavy rainfall and floods, while providing proper control of the weather for air transportation, national defense and health.[7] Weather control was still in its earliest stages of development, and yet the LRS saw it as a possible fix for several potential problems.

Even though Weather Bureau chief Francis Reichelderfer made another attempt to persuade him to drop the idea of the bill, Anderson continued to look for possible legislative models.[8] Triviz then suggested using the Atomic Energy Act of 1946 as a possible model because Anderson wanted to advance weather control by establishing a commission that would authorize licenses for conducting experiments. Perhaps the USWB chief could be an ex-officio member, and other members could be drawn from interested cabinet departments (Interior, Agriculture, Commerce, and perhaps State, but interestingly enough, *not* Defense). The bill would further indemnify contractors acting on behalf of the state and provide for international agreements. The Legislative Council Office agreed to look into it.[9]

The direct tie between weather control and water resources became clearer a few days later when Anderson, accompanied by other New Mexico leaders, advised the president's Water Resources Policy Commission to strictly control rainmakers before their activities got "out of hand." New Mexico governor Thomas J. Mabry concurred: his state needed moisture, but "indiscriminate seeding" could do more harm than good if damaging rains resulted.[10]

Anderson's pending bill got another boost when Drew Pearson published an overly enthusiastic column on the possibilities of weather control, which could turn the Far West into a "Garden of Eden" and other areas into deserts, totally disrupting the agriculture secretary's crop balance. Reporting that wheat growers in eastern Washington State had doubled their harvest by cloud seeding—a dubious, unsourced claim—Pearson suggested that rainmaking projects might make expensive irrigation projects unnecessary. Rainmaking could raise the water table in California's Central Valley, settle feuds over Colorado River water, and take water from one place and give it to another, all reasons why the "farsighted Senator Clinton Anderson" was proposing the legislation.[11]

With draft legislation proposing an independent Weather Control Commission in hand, Triviz discussed it with USWB, Air Force, Army Signal Corps, and Civil Aeronautics Board personnel. He was shocked to find them unsupportive. The USWB's take: "the most farsighted proposal ever submitted to Congress or possibly the biggest hoax." The Signal Corps representative suggested that a $500,000 research appropriation for the USWB made more sense than a large bureaucracy. One attendee referred to the draft bill as a "bourbon-eyed monster." Triviz had not heard comments like these from Langmuir. Stunned, he realized that if similar comments were uttered in an open committee hearing then Anderson's proposal would crumble.[12]

Conflicting recommendations left Triviz and Anderson in a tight spot. Reichelderfer, desperate to be taken seriously on a subject he openly opposed, sent a follow-up note.[13] New Mexico School of Mines president E. J. Workman urged Anderson not to use atom splitting as an analogy for weather control, and to keep an arm's length between his bill and the Atomic Energy Act. He also recommended pushing back against allegations that weather controllers were "invading the realm of Divine Providence," a claim that applied to any other method people use to control nature: irrigation, flood control, antibiotics, or even electric lights. Workman was especially concerned about the emphasis on military applications, because the connection between scientific development and weaponry might cause average citizens to become a mite paranoid, and security concerns could limit the number of people involved in weather control.[14] The Army Signal Corps argued against the bill in toto. How could scientists control something that they did not understand? And why protect the public from excessive weather control fees? It was not the government's responsibility to prevent people from spending their money for unproven purposes.[15] No nanny-state for the Army.

Commercial seeders worried that Anderson's legislation would restrict their ability to pursue rainmaking.[16] An especially hard-hitting appeal arrived from American meteorology's weather control entrepreneur extraordinaire, Irving Krick, who was lawyered up and ready to raise a ruckus. Under no circumstances would he support legislation curtailing the efforts of "pioneering scientists" engaging in rainmaking.[17]

As Anderson and his team laid the groundwork for introducing his Senate bill, Representative Thomas H. Werdel (R-California) volunteered to sponsor a companion House bill, and he introduced a resolution to create a joint congressional committee to study and investigate cloud nucleation.[18] (Table 3.1 lists all of the weather control-related bills.) The first round in the effort to regulate weather control at the federal level was imminent.

REGULATING WEATHER CONTROL—ROUND ONE (DECEMBER 1950)

In December 1950, Anderson introduced Senate Bill 4236—The Weather Control Act of 1951—to provide for the development and regulation of weather modification and control. It provided for programs assisting and fostering private- and government-conducted R&D, government control of experiments and operations to assure the common defense and national security, and administering policies and coordinating international arrangements to advance weather control.

Spurning Workman's advice to separate weather control from atomic energy, Anderson argued that it was not a sacrilege to control the weather any more than it was to split the atom. Water shortages in New York, dust storms in the 1930s, and deadly hurricanes illustrated weather's adverse effects. Insufficient rainfall killed crops, imperiled health, and shrank water supplies. Too much rainfall produced floods, washed away topsoil, injured and killed people and livestock, and damaged property. Weather was a national defense problem. Weather control could lead to profound changes in people's way of life, with far-reaching benefits to agriculture, industry, commerce, and the general welfare. Federal oversight was critically important to ensure its orderly development. The nation's population was moving west, water consumption was outstripping supply, and the nation had to increase its water resources.[19] Weather—specifically precipitation—and water were not just economically important; they were crucial to the nation's security during the Cold War.

Anderson received comments that can be sorted into three categories: supporting, opposing, and "squirrel food." The latter are the most entertaining and provide insight into citizens' uncensored thinking. We'll start with those. The Glare Research Institute of Chicago complained that GE's rainmaking endeavors were really about gaining illegal possession of its discovery of "contra blue rays." Just like every other rainmaking effort "dating back to the

TABLE 3.1. Weather Control Bills

Introduced	Bill No.	Title	Purpose
Sept. 1947	H.R. 4582		To investigate turning clouds into rain; $500,000 to Weather Bureau.
Dec. 1950	H. J. Res. 550	Cloud nucleation	To create a joint congressional committee to study and investigate cloud nucleation. Called for three members each from the House and Senate to determine the potential of cloud nucleation to induce or prevent precipitation, the need for licensing to prevent indiscriminate seeding that could lead to flooding and property damage, and the necessity for a federal agency to develop and control cloud nucleation techniques.
Dec. 1950	S. 4236	Weather Control Act of 1951	To provide for the development and regulation of weather modification and control.
Jan. 1951	S. 222	Weather Control Act	To provide for the development and regulation of weather modification and control.
Jan. 1951	S. 5	Water Production Bill	To provide for the production of water suitable for a variety of uses from saline waters as well as from the atmosphere; Interior Department as lead agency. Amended so that related research could be coordinated with the Defense Department or conducted jointly by Defense and Interior.
Feb. 1951	S. 798	Precipitation Research Bill	To authorize the Agriculture Secretary to conduct research related to the control of and provision of precipitation to areas that were low on water.
Jan. 1951	H.R. 1180		An act to facilitate the performance of research and development work by and on behalf of the Departments of the Army, the Navy, and the Air Force, and for other purposes.
July 1951	H.R. 4864	Weather Control Act	To provide for the development and regulation of methods of weather modification and control (Companion bill to S. 222).
Oct. 1951	S. 2225	Rainmaking Advisory Committee Bill	For an advisory committee to study and evaluate weather modification, authorize research and experiments by federal agencies, and protect citizens from exploitation and harmful and unwanted effects of rainmaking.
Mar. 1952	H.R. 7325	Rainmaking Advisory Committee Bill	Companion bill to S. 2225.
May 1952	H.R. 7785	Rainmaking Advisory Committee Bill	Companion bill to S. 2225

TABLE 3.1. (continued)

Introduced	Bill No.	Title	Purpose
Jan. 1953	H.R. 1064	Rainmaking Advisory Committee Bill	Reintroduction of H.R. 7325
Jan. 1953	S. 285	Rainmaking Advisory Committee Bill	Reintroduction of S. 2225
Jan. 1953	H.R. 1584	Weather modification programs	Authorized the Secretary of Commerce to provide for research, study, and safeguards related to weather control.
Feb. 1953	H.R. 2580		Created a committee to study and evaluate public and private experiments in weather modification (Advisory Committee on Weather Control).

Egyptians," the institute director wrote, this one was based on fraud.[20] Another correspondent claimed that the Russians had been developing cosmic rays that could influence the weather and the United States needed to surpass this effort.[21] The "weather weapon" theme cropped up repeatedly. One writer suggested that the United States should "starve out" the Soviet Union because the latter had a lot of land and a huge land army. He was so upset about newspaper articles reporting on this secret subject that he had contacted the CIA.[22] And then there were those who just thought weather control was crazy and a waste of money. From Illinois: "Trouble with you government men is that you wear too big hats for your brains."[23] Perhaps the first three correspondents had been spending too much time reading pulp science fiction. Contra blue rays? Cosmic rays? Death and destruction through drought? The last one? Not too impressed with government bureaucrats.

While the crank letters were being filed away, Anderson reached out to university-based atmospheric scientists. Common themes in their responses: the bill was too far-reaching; scientists did not know enough about weather control to regulate it; weather control was not equivalent to atomic energy; and, the real need was for more government-funded research. Views antagonistic to federal control also surfaced: leave licensing to the states and do not establish another bureaucratic agency like the Atomic Energy Commission.[24] One academic, who wanted to see coordinated research, provided a suitably pithy analogy: conducting several experiments at the same time in the same patch of air made "about as much sense as trying to find the separate effects of ten different fertilizers all at one time in one pot."[25] Support from atmospheric scientists? As pilots say, "No joy."

Who did support Anderson's bill? General Electric's scientific team, the same folks who had convinced him to draft it. The Weather Bureau and academic meteorologists supported funding for research, but not the creation of a new bureaucracy. The commercial seeders were none too happy either. Their solid opposition led to increasingly public and nasty exchanges with Anderson.

REGULATING SEEDING: WESTERN SHOWDOWN (JANUARY 1951)

As the Weather Control Act was being reintroduced in the new 82nd Congress, weather control supporters in the US West were organizing to oppose it. Their antistatist stand was quite clear: short of a war effort, regulation stymied private enterprise.[26]

The National Weather Improvement Association (NWIA), for example, was composed of cloud-seeding groups from throughout the West. Its mission was to raise and spend funds to oppose lawsuits that blamed artificial nucleation for "violence committed by nature," when, they argued, artificial nucleation was the way to *control* violent weather.[27] The NWIA's consensus opinion: the bill was premature and too complicated; the science was changing rapidly and the law would need to be modified; and cloud seeding endangered no one, therefore, the public did not need to be "protected" from it. Legislation should be simple to avoid hampering weather control's development.[28]

Irving Krick did not support the Weather Control Act either, as Anderson discovered while reading the *New Mexico Stockman*. But he would not have been surprised had he paid closer attention to the materials Krick had provided to him several weeks earlier. However, Krick had not explicitly revealed his opposition during discussions with Anderson, who wanted to break the story wide open. He thought the cattlemen were totally "selfish" for not wanting to see further studies on weather control and its implications. Anderson was convinced that New Mexico ranchers and cattlemen had seeded so much that they had prevented snow from falling in the Rio Grande watershed, shunting rainfall to their own ranches. He could think of no other reason why Krick and his counsel, T. R. Gillenwaters, would have persuaded them to lobby against the bill.[29]

Anderson had postponed introducing the Weather Control Act, he reminded the cattlemen, so they could gather more input. But Anderson saved his best barbs for Krick. He had wanted to believe that Krick was a "decent person" and had told Krick during his visit that bad stories were circulating about him. Anderson had given him the benefit of the doubt, even though some believed that "he is a faker and a grafter and that he just plays people for suckers." Anderson had given Krick a fair hearing and then watched Gillenwaters turn around and tell farmers that the bill was bad without giving Anderson a chance to reply.[30]

Anderson took his ire public in the *Santa Fe New Mexican*, charging that "haphazard rainmaking" was threatening the Southwest's economy. Overseeding caused by "uncontrolled and unobserved" rainmaking was leading to drought. Greedy individuals working to get a little extra moisture, Anderson railed, were destroying the normal rain pattern. He wanted New Mexico officials to enhance water resources for all citizens.[31] The *New Mexican* was not exactly a disinterested publication; its editor was Anderson supporter and Economic Development Commission chairman Robert McKinney. The newspaper supported the bill, arguing that scientists should be able to ascertain the validity of rainmaking "without interference from unqualified sorcerers."[32]

Picked up by United Press International, Anderson's January 1951 attack on cloud-seeding interests went national as he accused unregulated rainmakers of "robbing Peter to pay Paul." A cloud-seeding induced drought would not only harm agriculture, Anderson fumed, but also industry and "defense establishments" that had been placed in New Mexico at great federal expense. He wrote, "I think New Mexico had better watch its step."[33] Was New Mexico's junior senator threatening the state government back home?

Press reports indicated that north-central New Mexico's water situation was at its worst since the 1930s: snow levels were negligible, and wells between Taos and Santa Fe were running dry.[34] As the rhetoric heated up, accusations flew between Anderson and Krick and his supporters in the press and in the mail.

An angry Krick struck back, maintaining that his corporation's seeding in Colorado was not related to the Rio Grande Valley drought. Areas where he had been seeding were not drought-ridden.[35] His counsel, Gillenwaters, rather disingenuously claimed that he was "at a loss" to explain Anderson's statement that he was trying to turn farmers and ranchers in New Mexico against the Weather Control Act. He just thought it would be a waste of funds to set up this bureaucracy, which would slow down advances in this "revolutionary, scientific development."[36]

An incensed Anderson responded that not putting the federal government in control of weather modification would be a mistake. It controlled the highway system, but not because the states could not manage their own roads. Anderson had welcomed Krick to his office to discuss the bill and Krick had not stated his opposition. That Gillenwaters had "stabbed him in the back" did not sit well with Anderson, who would not have known of Krick and Gillenwaters's attempts to block his bill if friends in the livestock business had not tipped him off.[37]

While Anderson was promoting his bill, Krick was doing his best to defeat it. Speaking to "woolmen" in New Mexico, Krick argued that rainmaking should be viewed as a long-term strategy that would allow a wide variety of crops and livestock to be raised in semi-arid lands. At the time, Krick's company was seeding 100 million acres throughout the West, an area approximately the size of California.[38] No conflict of interest there . . .

A bill in the New Mexico legislature to allow rainmaking groups to organize as "benevolent, charitable, and scientific" nonprofit organizations exacerbated tensions. Influenced by cattlemen, the bill would allow groups of farmers, ranchers, and ski enthusiasts to form nonprofit groups to hire seeders and then write off their "charitable donations" on their taxes and evade responsibility if they were sued for property damage.[39]

The cattlemen turned out in force to support the Charitable Rainmaking bill, but when President E. J. Workman of the New Mexico Institute of Mining and Technology (formerly known as New Mexico School of Mines) took the stand to testify against it, those controlling the hearing cut him off—an event not lost on the local press.[40] To drive the point home, a state senator called on Workman and told him that New Mexico Tech's research appropriation would be in danger if he opposed the state's Charitable Rainmaking bill. An enraged Workman wanted to blast the legislators, but kept quiet until the research appropriation passed.[41]

Seeding groups opposed Anderson's bill, but those who felt victimized *by* seeding strongly supported it. Some constituents from north of Santa Fe thought they had lost crops to a cloudburst caused by seeding and then had been stricken by drought.[42] A rancher who lived just north in Colorado compared out-of-state commercial rainmakers to racketeers.[43] The Stockmen and Farmers group also endorsed state and federal regulation, as did the Farmington Chamber of Commerce.[44] An Albuquerque attorney thought it "perilous" to allow people to seed over large areas when the ultimate effects were unknown. He scorned Krick because his contracts were "fool proof," demanding upfront payment to cover expenses and profits, and building bonuses into the remuneration if significant rain resulted: a "heads I win, tails you lose" contract.[45]

Others supported the bill because individual state rules were clearly insufficient to deal with interstate seeding operations.[46] States were more easily "'manhandled' (lawyer-handled would be a better word) than the Congress," argued the *Alamogordo (New Mexico) News*'s publisher, who then mused that Russian scientists might try to get Pacific storms to dump their precipitation in the ocean, pushing the West into a drought. "That may be a fantastical idea," he wrote, "but nevertheless, it is something to think about along with the rest of the confusion and hysteria."[47]

Similarly, the California State Chamber of Commerce, while maintaining that weather control could be very beneficial, also claimed that only atomic energy presented the equivalent opportunity for disaster. Pointing to the secrecy surrounding some of these efforts, the chamber thought that the US military might not open up about weather control for a very long time.[48]

The prolegislation cohort included a wider range of interested groups than did those in the opposing camp. Those who thought they had lost property due to seeding and those who viewed seeding companies as racketeers, were considering the possible ways that weather control could be beneficial, and were looking for data to analyze had all come to the same conclusion: the federal government had a better chance of regulating weather control than did the states. It was time to get a law on the books.

As Anderson gathered constituent and stakeholder input, fellow senators introduced additional weather control–related bills, which muddied the legislative waters.[49] Multiple bills meant multiple committee hearings, but arguments between committee chairmen were finally settled by arranging a joint hearing on all of the relevant bills before representatives from three committees: Interstate and Foreign Commerce, Interior and Insular Affairs, and Agriculture.[50]

By late February 1951, Anderson began contacting potential witnesses. The strongly supportive GE group was willing to testify, but Langmuir's physician was so concerned about his testifying in a controversial setting that he forbade him to attend. Much to Anderson's relief, he later relented.[51] Anderson also sought witnesses from groups whose earlier unfortunate encounters with rainmaking, including the Arizona Cattle Growers Association and the Salt River Valley Water Users Association in Phoenix, had turned them into supporters of the bill.[52]

The military services, as usual, were not in total agreement, but at least they tried to coordinate their input. Their biggest complaint: they had been investigating artificial nucleation for several years, but doubted the tests' validity because of others' indiscriminate seeding. They needed legislation to protect their classified experiments from outside interference. However, the Weather Control Act was too expensive and too large. The military services wanted weather control work limited to military services and their contractors. Modifying the weather would be a criminal offense for anyone else, and the federal government should not pay for damages due to contractor negligence.[53]

The Testimony—Joint Hearings

Calling the subcommittee hearings to order, Anderson announced that members were primarily focused on the western states: California, the Rocky Mountain States, and states with salinity problems in irrigation water. He discussed rising temperatures across North America, particularly in the Rocky Mountain region, which was heavily irrigated. Rising temperatures and decreased precipitation had idled some hydroelectric plants. If rain- and snowfall continued to decrease, it would be important to reconsider irrigation projects dependent upon normal precipitation. Pointing to General Electric's research, Anderson held that it might be possible to "correct" this problem. He was not ready to say that weather control was possible, but he did think it was worth examining closely. Senator Lester C. Hunt (D-Wyoming), with a nod to the often-misattributed Twain quote, "Everyone talks about the weather, but nobody does anything about it," rejoined: "I hope we do not make a liar out of Mark Twain."[54]

Weather Bureau representatives testified first, coming under a withering attack from Anderson, who attempted to paint their agency as a recalcitrant organization unwilling to recognize Langmuir's brilliant work and Krick's notable rainfall enhancement successes in the West. The witnesses held firm: scientific knowledge of cloud processes, which needed to be understood before weather control would ever be economically viable, remained minimal. If the senators wanted to improve weather control's prospects, provide Weather Bureau scientists with sufficient funds to conduct their own experiments and to analyze the results obtained by commercial seeders. Until they could determine if weather control produced statistically significant precipitation differences, it made little sense to establish a Weather Control Commission.[55]

In contrast, the GE team was greeted with deference and a friendlier line of inquiry. It also took a completely different approach, focusing not on the lack of scientific underpinnings for weather control, but on the outsized role weather control could play in the nation as it was related to defense and economic prosperity, transcending state and national boundaries. Weather control, like atomic energy, required extensive research to fully develop economic and military applications. And lest the good senators miss the point, GE's man—Langmuir—*was* a Nobel Prize winner. They should place their bets on him. General Electric supported legislation that provided funding for its experimental work and kept others out of its way.[56]

A number of witnesses, including leaders from Interior, Agriculture, and their subordinate agencies, stressed the need for more water in the West: water for personal consumption, irrigation, and hydroelectric power. Already in 1951, witnesses noted that water was being "mined"; the aquifers from which it was being pumped would never be replenished. Weather control would bring needed precipitation to ensure a vibrant agricultural sector and eliminate the specter of drought.[57]

But some of the agencies' statements could have come straight from *Popular Science*. Not only could weather control bring gentle, non-eroding rain in just the right amount to a variety of crops; it could stop tornadoes, snuff out hurricanes, and prevent blizzards and floods.[58] Weather control was an all-in-one natural disaster prevention program.

The senators loved it. Discussing the possibility of moving precipitation from the rainy and snowy windward sides of mountains to the dry "rain shadow" side, Senator George Smathers (D-Florida) remarked, "You have to figure out how to guide that cloud sooner or later, to put a rudder on it, which also may have some possibilities."[59] Let's just say that the senators' collective grasp of basic atmospheric science was marginal. In the middle of the hearing, Wyoming senator Lester Hunt accused South Dakota of stealing his state's rainfall with its cloud-seeding effort, even though Wyoming is *upwind* from South Dakota, *not* downwind. Anderson jumped in to defuse the argument before it escalated: here was the problem with having no federal controls. States would be pointing fingers at each other, seeding to grab moisture for themselves, and "[destroying] the whole weather pattern" in the United States. If they had a *plan*, then everyone could get the weather they wanted.[60]

Others argued that cloud seeding and other potential weather control techniques would not respect political boundaries, hence the need for federal legislation that would address the entire United States, the only entity that could address the issue with other nations. Only the American state could act on everyone's behalf. Weather control was about more than water; it was about the weather, all weather, and it affected everyone.[61] But . . . it was mostly about water.

Vannevar Bush argued for more research and experimentation in a new agency that would spur competition. He would not give it to the "too skeptical" USWB because rapid progress came from "enthusiasts." When asked about weather control's military significance, Bush responded that he could see the possibilities, but "no clear pattern." Quizzed on the potential to modify the rainy season in Korea, Bush proposed using weather control in a different way: to alter crop-growing conditions. If weather could aid agriculture, then the United States could help countries overcome difficult climates and thereby "extend our favorable influence over the free world."[62]

Massachusetts Institute of Technology meteorologist Henry Houghton provided expert scientific testimony—so expert that it was way over the heads of his hearing room audience. He directly challenged Langmuir's assertion that it was possible to modify large-scale atmospheric processes, and then he made a pitch for funding and manpower for a long-term atmospheric research program and for protecting citizens from unscrupulous rainmakers. Qualified people in existing agencies could handle this—scrap the Weather Control Commission.[63]

The first witnesses had favored some action, even if it was scaled down from the bills being considered. But not all the witnesses supported federal intervention, and Irving Krick led off for the opposition. He testified that he had left the California Institute of Technology and taken the meteorology department staff with him to form a nonprofit, private research firm specializing in meteorology and related fields. (He did not say that Caltech president Lee DuBridge had eliminated the entire meteorology department to get rid of Krick and his vast array of consulting contracts that threatened his institution's reputation.)[64] Following preliminary studies, Krick had formed the Water Resources Development Corporation (WRDC) in 1950, to undertake large-scale practical applications of weather control. It had provided cloud-seeding services to wheat growers in eastern Washington and then had expanded into northeastern New Mexico and southern Colorado.[65] Krick and other seeders did not want any government interfering in their seeding programs.

The Military Viewpoint: Unexpected Opposition

Representing New Mexico, Anderson was focused on water resources. He was also concerned about the relationship between weather control and national defense, suggesting that the uncontrolled use of silver iodide could be more harmful to the nation's defense than "five atomic bombs." The danger would come from seeding that caused flooding rains in one area and massive drought in another. Silver iodide was more problematic than dry ice as a seeding material because the latter only worked locally, while the former could cover larger areas for a longer period.[66] Anderson thought his Weather Control Act was the answer; the military services did not.

All of the military branches opposed the weather control bills: they were overkill and other agencies might invade what had been, more or less, the military's exclusive turf. But they could not agree on what they *did* want, other than "modest controls" and basic licensing.[67]

"Modest controls" or not, the military services ultimately agreed that they needed restrictions on cloud modification to protect their experimental work, but not regulations that would hamper their efforts to explore additional weather control possibilities or mislead the public into thinking that weather control was imminent. The desired "simple" bill would require all seeding to meet regulations issued by a government agency, perhaps the Interstate Commerce Commission; create a board to provide expert advice to the regulatory agency, review the status of weather modification, and suggest additional legislation as necessary; require all seeders to provide information on their activities to the USWB; and require coordination of all seeding with DoD.[68]

Because of the hearing, the Research and Development Board (RDB) scheduled a meeting to discuss weather control's status with GE's Vincent Schaefer, MIT's Henry Houghton, and RDB legal counsel, J. C. Morrissey.[69] Schaefer and Houghton brought the RDB members up-to-date on seeding techniques and precipitation processes, respectively. Concerning military interests, Schaefer thought it might be possible to dissipate supercooled fogs, but not warm or ice fogs, and that large-scale control might be possible. Houghton thought it might be possible to punch holes in clouds and suppress or increase rain, and that large-scale control was *not* possible. He supported continued cloud physics research, as well as legislation that would aid research and development.[70]

Morrissey, presenting the legal briefing, opined that legislation would not stop litigation, and more litigation could impede research. The main question: "Who owns the clouds?" No one had an answer.[71] Morrissey was in touch with GE's counsel to ensure that DoD was aware of legal matters that could affect weather modification R&D. Indeed, GE had told him that it could no longer accept the risk of being involved in fieldwork due to potential liability, and that London underwriters had begun to insure against rainmaking damage.[72]

If the RDB and the military services were intent on using weather control as a weapon, they needed to continue receiving the latest information from GE and others doing this cutting-edge research.

REGULATING WEATHER CONTROL—ROUND TWO (JUNE 1951–DECEMBER 1952)

While the military services and DoD were attempting to hammer out a united position for Anderson, Senator Warren Magnuson (D-Washington)—whose state would soon be ground zero in the weather wars—was hearing from commercial seeding supporters, including the National Weather Improvement Association's Jim Wilson. The NWIA did not oppose federal legislation, but it thought that the earlier bills were too expensive and gave the federal government too much power. It wanted to finance and control its operations because, he said, "we believe it shows that old-fashioned virtues of character, integrity, and self-reliance have not gone completely out of style in the USA."[73] Having staked his claim on the American myth of self-sufficiency, Wilson neglected to say that NWIA hoped to continue its experiments without any reporting of, or interference with, its work.

As debate on Anderson's bill continued, additional weather control–related bills were introduced throughout 1951 in both houses of Congress. As if there were not enough bills floating around the Senate, Senator Francis Case (R-South Dakota) introduced S. 2225 (rainmaking advisory committee bill) in October 1951.[74] He claimed that experimentation in weather control had reached the stage where applications appeared practical, but the social, economic, political, and national security implications remained unknown. Therefore, it was important to fully explore artificial precipitation's potential uses.[75]

Unlike the earlier bills, Case's bill contained no controls. It would establish a temporary committee to examine the results of rainmaking experiments and research, and recommend permanent weather control legislation at the end of its two-year lifespan.[76] Once again, constituents, federal agencies, and the armed forces chimed in with their opinions. Constituent letters came in on both sides, but particularly from people who thought rainmakers were affecting the weather.[77] Some were convinced that atomic bomb tests had changed weather patterns in the Desert Southwest and that rainmaking efforts were worsening the situation.[78] An Illinois resident blamed heavy rains there on rainmakers in Arizona, Nevada, and New Mexico, whose cloud seeding was wafting east. He asked Case to "shut off" the rainmakers.[79] Anderson was getting letters inquiring about the Weather Control Act and complaining about "doctored weather" that had produced rain in the Ohio Valley and ruined crops.[80] A New Mexico farmer thought Irving Krick's rainmaking efforts were contributing to drought conditions and wanted him stopped, as did another hundred ranchers and farmers in Harding County in northeast New Mexico, who had signed the petition he had circulated against rainmaking.[81]

So not all letter writers were supportive. The Oregon Wheat Growers League had spent considerable time and money evaluating the efficacy of rainmaking and did not want to see any state or federal regulations until such evaluations had been completed.[82] Apparently league members had missed the point of S. 2225, which did just that.[83]

Defense-related organizations, still unable to present a united front on weather control, agreed that the proposed evaluation committee would need more than two years to do a thorough job. As a group, they recommended two years to construct an interim report, with a final report in five.[84] In early May 1952, S. 2225 was successfully reported out of committee and on its way to the full Senate, while its companion bills headed to the full House.[85]

Why did Case's rainmaking advisory committee bill survive the committee vetting process when other bills had not? Nonscientific and pseudoscientific rainmakers had been practicing weather control for a number of years, and the public was appropriately skeptical. Important questions remained unresolved: How often are atmospheric conditions conducive to successful weather modification? Will seeding produce economically viable amounts of rainfall? Another factor: rainmaking was becoming a big business. Millions of dollars were changing hands in attempts to bring water to semi-arid and arid regions. Although most meteorologists doubted the efficacy of weather modification, it was possible that it could have catastrophic effects and the public needed protection from *unnatural* weather disasters. Supporters included Vannevar Bush and the State Department, which acknowledged that international problems had not yet surfaced but that it would be a good idea to liaise with bordering countries since weather modification activities could not be contained within political boundaries.[86] Indemnity for damage was still a problem, although most of the experiments—fog dispersal, statistical evaluations—were not the least bit risky. However, hurricane modification could become a huge liability if seeding worsened its effects. Oregon's Senator Cordon pushed for indemnity for all government contractors.[87]

Passing out of committee does not mean coming up for a vote, and by fall 1952, S. 2225 was languishing. The military services were still whining about insufficient indemnity for private contractors, insufficient controls on private seeders, and insufficient time for the advisory committee to do its work. Fix those problems, and DoD would support the bill.[88] But despite the fixes, the bill died.[89]

REGULATING WEATHER CONTROL—ROUND THREE (JANUARY–AUGUST 1953)

As 1953 and the first session of the 83rd Congress got underway, the Rainmaking Advisory Committee bills were reintroduced as S. 285 and H.R. 1064.[90] Despite another Army attempt to modify the bill, S. 285 came up for a full vote, was passed in June, signed into law in August, and ultimately created the Advisory Committee on Weather Control (ACWC).[91] Senator Anderson had not gotten his desired Weather Control Commission, but he had secured an investigatory group that would analyze deliberate weather modification, judge its efficacy, and recommend legislation that would determine the future of weather control as a state tool.

President Dwight D. Eisenhower formally established the ACWC in late 1953, naming retired Navy captain Howard T. "Shorty" Orville, a consultant at Bendix Aviation Corporation in Maryland, as its chairman.[92] For the next several years, congressmen, the press, and public closely followed its work as hyped-up versions of weather control's promise continued to overshadow the day-to-day scientific grunt work to determine whether the promise would turn into reality.

WEATHER CONTROL IN THE PRESS . . . AND IN THE PUBLIC EYE

The press had been giving plenty of coverage to some of the wilder weather control efforts since the late nineteenth century, and the ACWC's visibility gave journalists additional opportunities to venture into a scientific field about which little was known, but which seemed to have possibilities for unlimited applications lurking just around the corner.

"U.S. to Do Something about the Weather" landed on the *New York Times*'s front page on December 10, 1953, but the much bigger story—that weather control could be used as a weapon—squeaked in on page 49 a day later. In the latter story, the *Times* reported that ACWC chairman Orville thought weather could be used as a weapon and that someday the United States might be able to induce flooding rains or prolonged droughts in the Soviet Union. He emphasized that America's Cold War nemesis, the USSR, could at most affect the weather in Alaska or western Canada. Most readers probably thought, "Well . . . who lives in Alaska or western Canada anyway?" However, Orville acknowledged that there was no scientific evidence supporting the possibility of modifying weather over long distances, contrary to Irving Langmuir's claims and military desires. He thought it would be possible to use seeding to prevent tornadoes and hail storms, but there was little scientific data to support these ideas.[93] Similar press reports prompted additional letters from concerned constituents and the continued promotion of "appropriate" weather control by Senators Anderson and Case.

Despite, or perhaps because of, the passage of the rainmaking advisory committee bill, Anderson continued to hear from constituents about the connection between drought and seeding. Others worried about atomic test explosions and the weather.[94] A resident of tiny Mountainair claimed that the folks at Kirkland Air Force Base in Albuquerque were sending out planes to bust clouds so that their radar systems would work properly. If this was a national defense issue, that was fine, but if not, then some other government department was "confiscating" the water he needed without compensation.[95] The Air Force's reaction: its planes flew *through* clouds; they did not do anything *to* clouds.[96]

Senator Case took his weather control ideas to the public. Attempting to show that weather control was becoming routine, he noted cloud-seeding programs all over the world: General Francisco Franco's government was trying to break the drought in Spain, and the French had "belatedly" tried cloud seeding on the "Reds in Indo-China . . . just ahead of the monsoons." The Australians were putting the largest part of their research budget toward rainmaking, hoping to convert a third of their desert into viable sheep-raising land. Case mused that if the Australians could bring water to the desert, one should consider what the United States might be able to do in its "short grass country." He often heard from people who thought it was "wrong to interfere with nature." To them he said: we drill wells and dam rivers, both of which interfere with Earth's natural course. The more we learn about nature, the better able we will be to control it. Or as ACWC member Lewis Douglas asked one of his neighbors who questioned the correctness of tampering with the weather: "Do you castrate any of your calves?"[97]

Rainmaking's supposed efficacy led to sensational newspaper accounts. The Associated Press (AP) reported that the United States might be able to "cause torrents of rain" by seeding, or cause severe drought that would dry up crops in the Soviet Union. The Russians, the article continued, could not so damage the United States because the weather "moves from West to East."

The source of this conjecture: none other than ACWC chairman Orville, who did not know if weather control would work or not, but his committee was trying to find out. The USWB's Harry Wexler, however, told the reporter that claims of marked increases in precipitation due to weather modification had not been verified.[98] And apparently neither the AP reporter nor Orville had recalled that the earth is spherical, and hence the eastern Soviet Union is *west* of the United States.

WEATHER CONTROL TANTALIZES THE PRESIDENT'S CABINET

The ACWC came up to speed quickly on potential state uses of weather control. Military uses had been under discussion since the late 1940s, but as money poured in and designer weather began to seem probable, civilian agencies began contemplating how they might advantageously use the weather. Members of the ACWC queried cabinet-level officials for suggestions. Agriculture had obvious uses for controlling adverse weather that often doomed thousands of acres of crops during the growing season. Others, such as the Bureau of Mines, seemed to be stretching for a reason to control the weather, but their ideas were no less engaging.

What is important here is that all of these departments and agencies were already in the "control of nature" business, so controlling the weather was just one more step along the continuum to total control of the natural world. The Department of Agriculture controlled fire ants, insects, and weeds using techniques that had been developed originally for military use . . . just like weather control.[99] Its Forest Service controlled fire and managed forests to ensure sustainable yields.[100] The public health arm of what had become Health, Education, and Welfare in 1953 had been heavily involved in eradicating malarial mosquitoes using DDT, giving rise to the Centers for Disease Control.[101] Interior and its Bureau of Reclamation (BuRec) had been controlling water resources since the early twentieth century.[102] The Fish and Wildlife Service sought to manage stocks of fish and game using maximum sustainable yield approaches, while the Bureau of Land Management similarly controlled grazing lands.[103] And the Defense Department? It was all about control.[104]

All of these agencies controlled something a bit different than the others, but when it came to weather control they desired the same thing: water. Water for crops, forests, and fire suppression (Agriculture/Forest Service); to fill reservoirs for irrigation and recreation (Interior/BuRec); to carry away effluent, but not so much that it flooded (HEW/Public Health); for adequate stream flow for fish and migratory water fowl (Fish and Wildlife Service); for grazing land and fire suppression (Bureau of Land Management); and for streamflow and hydroelectric power on the Columbia River (Bonneville Power Administration).[105]

Some agencies did not specifically ask for precipitation. Commerce, which included several agencies that dealt with transportation and industry, and Defense were keen on clearing fog and low clouds for aviation.[106] And the Bureau of Mines was interested in clearing particulate matter out of the air, perhaps by having it washed out by precipitation, or blown away by the wind.[107]

And the National Park Service was not interested in weather control at all unless it was going to impinge on the natural state of its parks. In that case, its officials wanted to make sure that any weather control efforts only affected land outside the parks.[108]

The newest agency in the weather control line-up was the National Science Foundation (NSF), whose mission was to support and encourage basic research and education in the sciences. National Science Foundation scientists argued that progress in weather control would depend on understanding the underlying physical mechanisms—knowledge that was currently unavailable. They advocated for more laboratory and field research, and expressed a willingness to sponsor conferences and symposia to advance weather control.[109] Perhaps this willingness to focus on basic research before jumping into unproven applications ultimately brought NSF to the forefront of civilian weather control.

Based on input from departments, bureaus, and agencies, government bureaucrats had wide-ranging ideas for using the weather to mitigate, solve, or prevent a whole host of social and economic problems. Regardless of the incoming data, the Advisory Committee on Weather Control could scarcely have concluded that experimental work should not continue. Press coverage that included science-fiction-like statements made by scientists involved in weather control research and applications had already led many to believe that operational weather control was within reach. During the Cold War, when people often felt at the mercy of powers beyond their control, the thought of people controlling something as seemingly uncontrollable as the weather would have been a strong motivator to continue. And it was. We'll pick up the rest of the ACWC story in chapter 5.

And what about Clinton P. Anderson's attempts to control the controllers? They did not prove so easily controlled—after all, how could one prove whether dry ice, silver iodide, water, salt, or the latest seeding material *du jour* had made a difference or not? And that inability to provide proof—on either side of this controversy—led each side to line up and take aim at the other. Anderson may have wanted to forestall the weather wars with his legislation, but he was too late. In states around the country, the battle lines had already formed.

CHAPTER 4

State Governments:
Averting "Weather Wars"

It shouldn't surprise anybody, in the near future, to see "Keep Off the Clouds! This means YOU!" signs displayed in prominent places on clouds everywhere.

"AMATEUR RAIN MAKERS," *Indianapolis Star* in the *Christian Science Monitor*[1]

If two adjacent landowners contract for different designer weather, what results? A weather war. Whether inter- or intrastate, one person's ideal weather could be a neighbor's economic disaster. In the early 1950s, commercial weather controllers started marketing their services to customers dreaming of more or less precipitation. Caught flatfooted, affected states scrambled to find a solution before dueling weather controllers created massive discontent among their residents. How could they keep all their citizens happy? They couldn't.

One might think most of these disputes would have taken place in the water-deficient US Desert Southwest or Intermountain West. No . . . the earliest rumblings of designer weather grievances came from opposite edges of the continent: New York City and Washington State. On the coasts or in between, state lawmakers, bureaucrats, and citizens spent the 1950s finding out how hard it was to determine where and when rain and snow were going to fall—or not.

WATER FOR THE BIG APPLE (1950-1951)

Facing a serious water shortage in early 1950, New York City leaders asked Irving Langmuir if cloud seeding could fill their reservoirs. Of course it could. Enthusiastically embracing rainmaking, the city's Department of Water Supply hired Langmuir's recommended meteorologist: Wallace E. Howell, a self-effacing researcher from Harvard's Blue Hill Meteorological Observatory. Howell, who held degrees from MIT and Harvard and who directed New

Hampshire's Mt. Washington Observatory, was a World War II veteran of the Air Weather Service, had served on the National Advisory Committee for Aeronautics Subcommittee investigating aircraft icing problems, and was associated with the US Weather Bureau's (USWB) Regional Forecasting Center.[2]

As Howell surveyed the city's water situation, the New York legislature introduced a bill freeing the city from rainmaking-related legal liabilities. The mayor of Albany, New York, had already attacked the plan, fearing it would steal water from his city. He was not alone; New England farmers were equally worried. Harvard meteorology professor Charles F. Brooks tried to allay those concerns by explaining the mechanics of rain-laden systems that move north along the Atlantic Seaboard. Since they picked up moisture en route, if watersheds feeding New York City's reservoirs received an extra 0.10 inch of rain, that would likely reduce rain across New England by only 0.001 inch—a miniscule amount. Water vapor picked up from the Atlantic Ocean would bring normal rainfall to points north of the city, seeding or no seeding.[3]

Complaints aside, the plan continued. Five planes for aerial seeding were made available: four from the police and one from American Airlines.[4] An advising team composed of local physics professors joined the effort, and USWB chief Francis Reichelderfer promptly reminded them that it was impossible to evaluate rainmaking experiments without first carefully analyzing the synoptic weather situation.[5]

Filling the reservoirs with man-made rain would not be cheap. Howell was charging $100/day ($1,000 in 2015 dollars), and $50,000 ($500,000 in 2015 dollars) had been appropriated to fund the first six months. The *New York Times* reported it would be the "first time rainmaking has been attempted scientifically for practical purposes," discounting all of the military activity already underway. Besides boosting the water supply, Howell hoped to gather data applicable to other regions. Determining whether rain had stemmed from natural or man-made causes was difficult, he cautioned, and rain would only fall during appropriate weather conditions.[6] The inability to determine whether a particular cloud-seeding attempt had triggered resulting precipitation continued to dog weather controllers throughout the century.

A *New York Times* editorial pointed out that Howell could not guarantee rain, but the appropriated funds were small compared to a potential return of water worth millions of dollars. Langmuir claimed that rainmaking depended upon both the synoptic situation and the number of ice nuclei, natural or artificial, in the clouds, the *Times* continued, but the USWB was only interested in the synoptic situation.[7] That comment did not sit well with the thin-skinned Reichelderfer, who challenged this "criticism" as being typical of rainmakers who "misconstrue the bureau's desire to obtain facts" before forming conclusions. It would be more accurate, he argued, to say rainmakers ignored the synoptic situation entirely. The USWB, in contrast, had closely considered probabilistic and statistical work, and it made every effort not to allow "wishful thinking" to guide its conclusions on weather control. The controversy surrounding rainmaking would disappear if everyone published well-documented reports. Was weather control possible? Certainly. It just was not economically viable.[8] City bureaucrats didn't seem to care if scientific evidence was lacking; only the possibility of success mattered.

Howell planned to use both ground and air seeding, and he hoped to station observers in the Catskill watershed to report on rainfall.[9] Although 1950s-era weather radar was terrible compared to today's easily accessible NEXRAD images, it was better than nothing. The Air Force loaned one to the project so meteorologists could determine where rain was falling. They could check the seeding plane's location against the radar image to determine if the rain fell before or after the seeding started.[10] As was typical for weather control experiments at the time, there is no evidence that Howell installed a rain gauge network to record rainfall on the ground. Therefore, hard data for verifying the efficacy of his seeding efforts were unavailable.

Lawsuits landed on city desks before the first seeds flew. Property owners and civic associations in Ulster County, the northernmost county in the New York City metro area, claimed the city was "about to become engaged in a dangerous and unusual experiment." Seeding could trigger an unstoppable chain reaction that would lead to a stream-flooding, people-drowning, property-destroying deluge. Mayor William O'Dwyer's dry response: "Somebody does not want it to rain, I take it."[11] "Somebody" had probably been reading too many accounts of "unstoppable chain reactions" associated with atomic bombs. Worries about nuclear bombs and weather control were common during the 1950s. Similarly, in the absence of solid scientific explanations for resulting rain, answered prayers might be a possibility. When rain started falling *before* the seeding started, an itinerant Baptist preacher from Texas claimed his prayers had been responsible for the precipitation, and asked the city for a $7,000 donation. New York governor Thomas Dewey declined to comment.[12] Wise decision.

While city crews worked to staunch leaks in the water system, other workers prepared the dry ice. Police airplanes awaited Howell's signal to fly, but rain was falling, and officials did not want to enhance rainfall and cause stream flooding.[13] As GE's Vincent Schaefer was telling the Royal Meteorological Society in London, New Yorkers should not count on cloud seeding to fill up their reservoirs. Seeding worked best when rain was imminent, and he was concerned about sensationalized press reports. While praising the city's efforts, Schaefer warned that neither Howell nor anyone else could work miracles.[14]

In early April, Howell directed the first dry ice–seeding run, but he could not determine if the resulting unseasonable snow was natural or artificial.[15] Natural or not, unhappy residents jammed city hall's phone lines to complain about "Howell's Snow." Thrilled city officials credited seeding for the additional moisture.[16] Days later, heavy rains sent water over the banks of a city reservoir, but officials remained unsure of its cause.[17] Once the water receded, Howell began seeding with silver iodide over the Catskill watershed.

Howell declined to take credit, even though a cursory check of surrounding watersheds found more rain in seeded areas. He suggested comparative studies of natural and artificial precipitation.[18] But how to tell the difference? The drops looked the same, acted the same, and did not contain markers allowing them to be sorted. Not even the weather controllers knew for sure.

As April turned to May, the rainmaking team used ground generators to send billowing clouds of silver iodide smoke into the clouds, which yielded little rain.[19] But by mid-month, weekenders and farmers downwind from the seeders were howling that seed-triggered rain was spoiling their activities. Annoyed by the complaints, Howell and the meteorologist-in-charge of the local USWB office argued that the weather was entirely natural. They would not know for months if seeding efforts had boosted rainfall. Unhelpfully, the Weather Bureau official added, "Only God can make a cloud."[20] Natural or not, heavy rainfall coupled with water conservation efforts filled reservoirs. As Memorial Day approached, residents besieged city hall with requests to stop making rain lest it spoil outdoor fun. But Howell only seeded when conditions were ripe for rain, so if the rains were going to come, they would regardless of cloud seeding.[21]

By early June, the reservoirs were almost full and consumption was dropping. The rainmaking team seeded from above and below the clouds, and heavy rain fell over the watershed. Howell still refrained from taking credit. Despite his demurs, those whose holiday weekend plans had been washed out, and businesses that had had fewer customers than expected, bombarded city hall with protests. Officials continued to deflect criticism: there was no proof that seeding had contributed to the heavy rains.[22] And therein lay the beauty of cloud seeding: by not taking credit for the "good" results, one could escape taking the blame for the "not so good" results as well.

Soon the reservoirs were not just full—they were overflowing. Howell temporarily stopped seeding. The Dutchess County agricultural agent pleaded with the governor's office: halt the seeding! Albany was being hit by drought, and the agent suggested shifting the seeding projects further north.[23] But two weeks later, the water levels had receded and Howell prepared the ground generators for more seeding, which continued into early July.[24] Meanwhile, seventy-five miles north of the city, the Sullivan County Board of Supervisors passed a resolution calling the city's rainmaking efforts a "public nuisance," hinting they might request a grand jury inquiry. They urged city officials to stop rainmaking to "avoid future ill-feelings and legal activity."[25]

Howell's contract ended in August. Should city officials extend it or not? Mayor O'Dwyer was extremely enthusiastic about continuing, even though no one, including Howell, had claimed that seeding was responsible for filling the reservoirs. Rainfall was 8 percent above normal for the calendar year, and for the first time in fifteen years, the reservoir waterline had risen between June 1 and July 16.[26] Because they wanted data from all four seasons, the city extended the contract for another six months. The *New York Times* headline: "Rain-maker Gets Another 6 Months."[27] It sounded as if the city had tossed Howell into Rikers Island jail and left him there; those northern counties would have been cheering.

Farmers had mixed feeling about extending the rainmaking project. Some farmers in the southeastern counties thought seeding had helped their crops, and others thought Howell had "dug a watery graveyard" for their hay and caused a fungus ruining their apples. The Dutchess County Farm Bureau agent argued that control of rainmaking should rest with the state because it was not fair for the city to induce rainfall during the haying season. All the farmers were convinced that Howell was changing the weather. One Hyde Park–area farmer reported that he had seen "freakish clouds" appear in the middle of the day. "The sun would be shining and then a funny-looking cloud would come along and it would rain for about ten minutes—just long enough to wash the spray off the apple trees. I know some people," he continued, "who would shoot Dr. Howell on sight."[28] The farmer's response was common. No one had proof that seeding changed the weather, and in its absence people would have grumbled about the inopportune rain or complained about the forecast, but cloud seeding led to the default assumption that it caused undesirable weather, wet or dry.

New York City continued to support the seeding program, but farm organizations were lobbying New York governor Thomas Dewey to regulate rainmaking. They argued that if the government could regulate radio's "ether waves" and aviation airways, then it ought to regulate the wringing of precipitation from clouds. Five days later, the state assembly introduced a bill to do so.[29]

Meanwhile, disgruntled property owners continued to file lawsuits against the city. In February 1951, eighty property owners in a resort area of Delaware County (about 150 miles north of the city) filed a $1 million lawsuit claiming damage from induced rainfall.[30] This was followed by an almost $300,000 lawsuit filed by residents of upstate New York. By the summer, additional residents—including the operator of a "seal college"—claimed artificially induced rainfall flooded a local creek, damaging their property.[31]

By late 1951, New York City appropriated funds to determine whether the previous year's rains had occurred naturally. City experts thought the 1950 Thanksgiving storm had blown in from the Ohio Valley (it had) and was not due to seeding (it wasn't). Damage claims due to weather modification were "so new" they thought it prudent to conduct an intensive investigation and escape a legal hook.[32] With the end-of-year deadline upon them, lawsuits flooded New York City's Corporation Counsel office. Totaling $1.5 million, over one hundred summonses and complaints named city officials and Wallace Howell as defendants. They charged Howell with "trespassing" on their land with his recklessly produced rain that had damaged their property.[33]

Once again, the *New York Times* editorial board chimed in, opining that the city had enhanced problems more than rain by hiring Howell. The USWB continued to maintain that rainmaking was a "dubious procedure," but conceded it was possible to locally alter clouds. The American Meteorological Society (AMS) had warned in its policy statement that there was "no present scientific basis for the belief that we now possess the ability to modify or control the weather and climate of a major portion of the country," but it still supported cloud physics research. The editorial board concluded that cloud seeding still held out the most hope of enhancing precipitation of all the other techniques that had been tried since the mid-nineteenth century.[34]

But in 1953, when New York City once again faced a water shortage, the city decided against rainmaking and in favor of conservation, estimating that the city lost 100 million gallons of water each day due to leaky pipes and faucets. It made more sense to repair the leaks.[35] The decision was influenced by the state assembly's passage of a bill giving the Water Power and Control Commission jurisdiction over artificial rainmaking.[36] The legal entanglements involved with "making rain" outweighed the benefits, particularly when that expensive rain was never making it into residents' homes. The added benefit: repairing leaks was not controversial.

Although rainmakers, Howell included, were not necessarily claiming credit for making or enhancing rain, regulating weather controllers at the state or local level became contentious during the 1950s. What kind of heartburn was rainmaking creating on the other side of the continent? Normally damp Washington State provides another view of competing designer weather.

DAMP IF YOU DO . . . DAMP IF YOU DON'T: DESIGNER WEATHER IN THE PACIFIC NORTHWEST (1951–1952)

When drought struck the Pacific Northwest in the early 1950s, it menaced hydroelectric power needed by aluminum smelters, not urban residents. Aluminum was a Cold War strategic asset, so the federal government moved in to help with cloud seeding. The Department of the Interior announced it would seed 61 percent of the watershed that fed the reservoir behind the Columbia River's Bonneville Dam. Although this was not a "sure cure," Interior hoped for success. Some aluminum production lines had already shut down for lack of hydropower. Industrialist Henry J. Kaiser thought he would be able to continue operating his Pacific Northwest aluminum facilities, but only if Washington's residents made "every sacrifice necessary."[37] The citizenry would need to skimp on water and power to maintain aluminum production, even though Kaiser would have been hard-pressed to find cheaper power elsewhere.

In addition to boosting water to power Kaiser's aluminum plants, weather modifiers had contracts with farmers and with power and lumber companies throughout Washington. University of Washington meteorologist Philemon E. "Phil" Church was worried about the increasing presence of rainmakers, and that adverse public reaction might affect both the university and its meteorology department. In a letter to his dean, he laid out several concerns. Only the rainmaking firms knew the time and location of seeding, materials used, and results. Consequently, he and his colleagues could not verify them. Some of the companies' claims were clearly false, and people were being "fleeced." Church held that while two of the firms were extremely professional, two were unethical. With no laws to regulate their activities, there was no control over rainmakers. The meteorology department's faculty members felt obliged to protect the state's citizens, who were currently shelling out approximately $150,000 per year ($1.5 million in 2015 dollars) to rainmakers operating on both sides of the Cascade Mountains, which run north-south through the western half of the state and create the dry "rain shadow" on the eastern side.[38] Church's letter landed in the hands of University of Washington president Raymond B. Allen, who forwarded most of the contents to Washington governor Arthur B.

Langlie. The university was willing to help, but had no funding to conduct the necessary research.[39] Although the letter did not include an "ask," President Allen implied his meteorologists would be on the job in no time if funds were forthcoming.

Within two months, the state legislature's Subcommittee on Natural Resources announced hearings on artificial cloud stimulation. The first hearing, held in December 1951, attracted academics, commercial cloud seeders, and state officials. Testifying on behalf of Irving Krick's Water Resources Development Corporation (WRDC), then seeding for wheat farmers in semi-arid eastern Washington, was the aptly named T. H. Hazzard. After discussing the technical details of weather modification, he told the legislators that he and his team were not "rainmakers," they were "rain increasers" who produced more rain than would have fallen naturally. The legislators asked: if more rain falls on an area, what happens to areas downwind? No problem, Hazzard told them. Doubling the rainfall in the target area would still leave 99 percent of the rainfall for those in the storm system's path. Why? Because natural rainfall was inefficient, and that allowed his firm to "operate" on the same system as it moved across the continent.

What about people who did not want extra rainfall? Everyone in the West needed more water. At the very least, additional rain would raise the water table, benefiting everyone—a claim like the one Frederick Newell had made in the early twentieth century that irrigation would be good for every state. The WRDC team was increasing rain; people might get a little wetter if it were seeding. Hazzard also claimed seeding clouds led to gentler, less wildly uncontrolled, less erosive, and thus more valuable rainfall.[40] The "kinder, gentler weather phenomena" argument was virtually identical to the one proffered in the nineteenth century. Modifying rainfall was not just about more rain; it was about the *right kind* of rain in the *right place*.

Hazzard understood that people were worried about the 300 million acres of land affected by seeding. Some in Washington State thought the firm was going to bring in clouds and open them up right over ripening cherries, causing them to split, and ruining the harvest. He wished he and his team had that much control over nature; they could do so much good. When people understood what they were doing and why, they did not find much opposition. When legislators mentioned grain growers who opposed his activities, Hazzard was dumbfounded. Why would they protest cloud seeding? Every grain grower in eastern Washington needed more rain.

When queried about a recently completed Colorado A&M (now Colorado State University) cloud-seeding study that countered the WRDC's opinions,

Hazzard admitted that he had not read the report, but he had heard of it and its author's claims that insufficient data existed to develop a firm conclusion about rainmaking's efficacy. The WRDC agreed: there was *no proof* that its weather modification techniques enhanced rainfall. Its teams had noticed "very unusual things" occurring after their operations and assumed seeding was the cause. However, it was impossible to make a judgment after operating just one year in Washington; the WRDC would need at least five years to make a final determination.[41]

The legislators also wanted to know why rainmaking had stopped in Europe. Hazzard summed up the reason in one word: lawsuits. Cities, industry, and agriculture were closer together there, and it was impossible to create weather that met everyone's needs simultaneously. For that reason, the WRDC did not operate in the eastern United States.[42] And therein lay the crux of the problem: people wanted different kinds of weather, and if it worked, some of them were not going to get the weather they wanted. Where Hazzard erred, or deluded himself, was in thinking that this would be a problem only in heavily populated areas.

Support for rainmakers came from a predictable quarter: the wheat farmers who had hired them. The farmers contracted with rainmakers because they needed water and would take necessary steps to get it. The cost was $0.15 per acre, or about $500 (about $5,000 in 2015) for the average farmer. The average annual rainfall was only 7.5 inches, and the enhanced rainfall meant a significantly better wheat harvest. Other than grass for cattle, wheat was the only viable crop in eastern Washington. The wheat farmers agreed that rainmaking needed evaluation and control. They were not the only agriculturalists who wanted to know whether it was economically effective. Orchardists, including the cherry growers, wanted to control rainfall so that it did not fall during blossom time (knocking the blooms from the trees) or during harvest time (causing the cherries to split). It was impossible to control something like precipitation when so little was known about it. Cloud seeding was still in the "little known" stage, but some growers thought that all seeding should be prohibited between April 1 and November 1 to reduce possible crop damage.[43]

Despite its uncertainties, one of the legislators wondered if rainmaking could replace reclamation projects. Would the money being spent on the Columbia (River) Basin reclamation projects be wasted if rainmaking could bring water to where people needed it, when they needed it?[44] Having reached no conclusions, the subcommittee adjourned.

In spring 1952, the subcommittee held another hearing that included a different cohort of interest groups, including cherry growers, foresters, and meteorologists.[45] Orchardists viewed rainmaking differently than did their wheat-growing colleagues. Cherries and other fruit crops suffered serious damage if rains came at inopportune times. They had no control over natural rain and could live with that, but rain caused by seeders was a problem. Since the government had put in the irrigation system they were using and paying for, the farmers needed to harvest a decent crop to earn the money to pay for it. One orchardist had lost $35,000 worth of cherries when one of Irving Krick's WRDC seeding operations spread rain outside the target area (or so he claimed). Other farmers had lost crops as well: apple growers had been unable to harvest, bean growers suffered large losses when the harvested beans rotted in their sacks, and alfalfa-seed losses had totaled several million dollars. Orchardists acknowledged others' need for water, but everyone needed to be responsible for their actions. If irrigators let water overflow into others' fields, they were liable for the damage. If the wheat growers wanted to experiment with rainmaking, that was fine. They should not do it, however, in a way that could financially harm farmers for whom rain was a problem.[46]

Those involved with lumber and forestry thought controlling rainfall would greatly benefit fire-protection and logging industries. For example, in 1951, Washington State's Forestry Division had had to close down logging for almost a month because of excessively dry conditions that exacerbated fire dangers. Both lumbermen and foresters wanted more data from controlled tests, and sufficient regulation to keep the air from being "polluted with artificial seeds."[47]

Representing meteorologists and the AMS's Puget Sound Chapter, Phil Church testified that atmospheric scientists wanted a full accounting of weather modification activities: what rainmakers were doing, what materials they were using, and how, when, and where they were using them.[48]

Unlike the earlier hearing, this one addressed the topic of *over*seeding: using large amounts of silver iodide seeds to dry up clouds and prevent rain. The cherry growers were interested in overseeding because they could achieve higher profits if they could prevent rain during blossom and harvest times. Would proposed legislation allow them to overseed?[49]

The cloud seeders also participated. Water Resources Development Corporation's Robert Elliott, formerly of Caltech's meteorology department, argued that any legislation had to be based on extant knowledge about weather control, and there was considerable disagreement within the scientific community. "Our exact knowledge," Elliott conceded, "is so nebulous at this time that . . . I don't think that any of us could propose any clearly defined legislation."[50]

The hearings settled nothing and by late June 1952, a full-fledged "weather war" was underway between grain and cherry growers. In a letter to Phil Church, State Representative Wilbur G. Hallauer only half-jokingly wrote that the cherry growers were probably working hard at overseeding so that no rain would fall until clouds reached Montana or points east. Convinced that professional meteorologists needed to be involved in cleaning up this mess, he sought suggestions.[51] Church recommended that weather control operators be licensed, exhibit a high degree of professional competency, provide transparency when dealing with the public, and supply data for independent evaluations. Church, however, did not see a way to legislate happiness for the conflicting sides:

> If the cherry growers succeeded in getting legislation to prevent rain in a period, then the railroads might want a provision such that they could prevent snow in the winter to the detriment of irrigation and power interests, housewives would want legislation such that they could be guaranteed no rain on Monday from 10:00 am to 2:00 pm so the washing could dry, and office workers would seek legislation to aid them in having dry streets when they drive to and from work.[52]

And back at the weather wars? Krick and the WRDC, working for the wheat farmers, had been seeding for rain, and their rainmaking competitor, Jack Hubbard, had been overseeding to keep cherries dry. Subsequent heavy rainfall caused crop losses for both groups.[53] The AMS's Puget Sound Chapter stayed out of this dispute but told Governor Langlie that seeders' activities should be public knowledge; its members assured him that they would be happy to provide advice on future legislation.[54] That legislation remained stalled for another year, and another, and another. We'll catch up with the State of Washington's attempts to regulate the weather controllers a little later.

STATES IN THE MIDDLE: EXPERIMENTS AND REGULATIONS (1950–1953)

Taking a cue from New York City, states in the Desert Southwest and Intermountain West, perennially short on water, were considering their own large-scale rainmaking operations by early 1950. With the East suffering from drought, westerners thought it might be easier to get more funding from the Bureau of Reclamation (BuRec), which had been responsible for water supplies west of the Mississippi River since 1902.[55] Several states explored the possibilities of systematic rainmaking activities as well as the wisdom of passing legislation to regulate cloud-seeding activities. Legislation often included establishing a board or committee composed of academics, agriculturalists, and industry interests to investigate and/or mediate disputes over cloud seeding.

By mid-1950, New Mexico became the first state to launch its own rainmaking study. Governor Mabry formally approved the newly established Economic Development Commission's decision to have the New Mexico School of Mines (soon to become the New Mexico Institute of Mining and Technology, or NM Tech) in the tiny high desert town of Socorro conduct a four-month-long study of all aspects of rainmaking. Irving Langmuir, in New Mexico with Project Cirrus, agreed to collaborate. Experimentation was not the only point of contention. Many farmers and ranchers had decided to take rainmaking into their own hands. Langmuir was concerned that such "indiscriminate rain induction" could cause severe weather elsewhere (see chapter 2) and might have contributed to drought by inducing storms in Texas that had prevented the normal northwesterly movement of moisture from the Gulf of Mexico.[56]

Analyzing the resulting data, School of Mines scientists thought the experiments had yielded one snowstorm and two cloudbursts. They did not, however, make a "positive claim of results." The snowstorm had taken place twenty-five minutes after silver iodide generators had started spewing out seeds into mountains northeast of Santa Fe, and snow only occurred in the clouds affected by silver iodide smoke. The cloudbursts had resulted from earlier tests, which had used several hundred pounds of ammonia as the seeding agent.[57] Once again, the results were not definitive.

Wyoming also decided to act on weather control. In February 1951, the state legislature passed an act declaring sovereignty over atmospheric moisture overlaying the state, encouraging rainmaking experiments and their evaluation, requiring licenses for seeding experiments, and establishing a rainmaking commission comprised of the state engineer, the commission of agriculture, and the University of Wyoming president. The annual licensing fee: $25.[58] Later that year, Wyoming became the first state to grant a cloud-seeding license.[59]

In fall 1951, the Wyoming Agricultural Experiment Station completed five months of analyzing commercial seeding results. Some experiments had produced significant increases, but there was no proof that large-scale ground seeding operations had produced significant precipitation. As was common, the results had fallen within natural variation. Areas receiving little rainfall on average often experience large inter-annual rainfall differences, making statistical analyses difficult. The research team could not reach a firm conclusion on seeding efficacy.[60]

Meanwhile, residents of five Colorado mountain towns complained that cloud seeding was causing mineshaft cave-ins that threatened their livelihoods. Miners presented a petition to Colorado governor Daniel I. J. Thornton, claiming that they had been buried under record amounts of snow since seeding had commenced in fall 1950. Following a second complaint about excessive snowfall, Thornton asked his state's Weather Control Board to examine the allegations. Businesses in the Estes Park resort area sixty miles northwest of Denver complained that cloud seeding was ruining its tourist trade.[61] By late June, the Weather Research Association—while denying any responsibility—turned off its eighteen ground generators in southeastern Colorado because too much rain, some with hail, had fallen.

Colorado also passed a weather control act in 1951 to assist private and public weather modification R&D, and protect people and property from associated damage. The act defined weather modification as changes to weather phenomena and air masses: temperature, wind velocity, precipitation, or fog. The related advisory committee included academics, as well as representatives from state agencies, utilities, agriculture, and industries that would likely be affected by weather control.[62]

Farther west, California and Oregon lawmakers were also working on cloud-seeding legislation. California's bill called for regulating and licensing of "interference or attempts to interfere" with natural precipitation processes: "public interest, health, safety, welfare, and necessity" required scientific experimentation in weather modification, which should be encouraged to help conserve and develop state water resources.[63] By late 1952, Oregon's legislation also promoted public health, safety, and welfare by licensing, regulating, and controlling artificial techniques of inducing precipitation. Specifically, it decreed that no one could participate in weather modification activity without a license. Those who met licensing requirements and engaged in weather control had to submit reports and pay appropriate fees.[64] In early 1953, Oregon introduced additional legislation to regulate weather modification activities, appropriate funds, and provide for controlled experiments and penalties. A State Weather Modification Board would issue licenses and supervise weather control activities. Oregon State College's (now University) R. T. Beaumont advised legislators that any weather control regulations should be the same as Washington's. Experimental work done on the Columbia River, the boundary between the states, would affect both of them; consistent rules were necessary.[65]

Assessing Oregon's situation since 1949, Beaumont noted nine different projects that included rainmaking, hail suppression, and fog dispersal. Public opinion ranged from complete approval to bitter resentment. Southwestern Oregon's Jackson County had narrowly rejected a bill to outlaw cloud seeding because scientific evidence did not support either side. Beaumont opined, "Cloud seeders have better press agents than the groups performing impartial evaluations." He recommended that Oregon undertake an impartial experiment, making sure that the chosen test area would not be impinged upon by other cloud-seeding operations.[66]

Looking back east, Massachusetts established a Weather Amendment Board consisting of commissioners of agriculture, public health, and conservation, to certify and approve the actions of would-be weather modifiers. Public hearings on the operations had to be announced at least forty-eight hours in advance. The bill clarified that heating or moving air to preclude crop damage, e.g., using smudge pots or large fans, was not weather modification.[67]

Whether on the Eastern Seaboard or the West Coast, or in the Intermountain West or Desert Southwest, states tried to keep the possibility of weather control alive while keeping controllers on a relatively short leash. One group wanted to see that leash either become longer or disappear altogether: the commercial seeders.

COMMERCIAL SEEDING AND THE STATES

As the push for state and federal legislation became serious in 1951, consulting meteorologists interested in weather control organized the Artificial Precipitation Operators Association. They did not want their operations restricted, nor did they want people to be victimized by unethical rainmakers. Others engaged in cloud seeding, including some former crop dusters, were ethical; they just lacked basic scientific understanding. If association members could provide them a solid scientific background and help them evaluate their results, the crop dusters-cum-cloud seeders could bring their work up to higher standards. Association members wanted to "clean up the situation" before a government agency did it for them.[68]

But while those commercial seeders were on the lookout for less-qualified individuals infiltrating their ranks, some government entities were eyeing consulting meteorologists active in cloud seeding. The primary target: Irving Krick and his Water Resources Development Corporation.

In 1951, Krick came under increasing scrutiny in New Mexico, where he had been hired to break the drought. The *Albuquerque Journal* reported that the drought was continuing and Krick had not produced promised rain. Krick responded that he and his men were not "miracle men": they could not create rain, only enhance it under the right conditions.[69] New Mexico's farmers had made a significant investment in Krick and his weather modification operation, more than $115,000 (about $1 million in 2015 dollars) to alleviate the drought. Water supplies were low, and fires were a problem too.[70] Krick claimed that half of the rain that had fallen in the state was due to his work; the entire state was under contract to his firm. "We've got New Mexico pretty well taken care of," Krick said, claiming the drought would be over by July.[71] Or not. In late May, farmers in the San Luis Valley pulled out because rain was not falling. Krick argued that it was not his fault that there were no clouds to seed.[72] By late July, Krick declared the drought "broken," but New Mexico had a way to go before being back to normal. The USWB took a less-positive view: there was little hope that the state would meet its normal rainfall average unless it was hit by a series of cloudbursts.[73] No one in New Mexico, save Krick, believed the drought was over.

New Mexico's water problems aside, Senator Clinton Anderson was concerned that Krick may have caused flooding in Oklahoma. Was he operating there or not?[74] Krick was not and denied any connection between his seeding projects and flooding in either Oklahoma or Kansas. Indeed, the idea was "laughable."[75] But former Krick acolyte Robert D. Elliott, who had abandoned the WRDC to join North American Weather Consultants, a cloud-seeding firm headed by meteorologist Eugene Bollay, was concerned that Krick's activities would give all cloud seeding a "black eye." Elliott told Anderson that he (Elliott) and Bollay were using a precise, predictable technique, not a "shotgun method," and it would bring water of economic value.[76] Elliott repeated his message to Irving Langmuir, claiming that he and Bollay could find no downwind effects more than fifty miles away from their seeding generators and that they were using 1/100th the amount of seeding material they had used in the past and that many others were still using.[77]

Krick's methods were also being eyed in Idaho, where T. H. Hazzard (who we met in Washington State) spoke to Idaho Reclamation officials about rainmaking. Explaining that airborne dry ice could not be evenly distributed, he touted Krick's new seeding technique that would overcome these problems. After giving his "we are rain enhancers" pitch, Hazzard claimed Krick's group was operating on over 250 million acres in the United States, South America, Mexico, and South Africa. In the United States, it had covered most of the central western states. Hazzard estimated Krick's teams could increase rainfall by 50 percent to 75 percent, but they did not seed in areas where they could accomplish very little. They had increased rainfall in eastern Washington State by 406 percent—a phenomenal amount. (What was really phenomenal was that they thought their calculation was scientifically significant: 406 percent?)[78]

Hazzard maintained that the WRDC could provide more water at less expense. It was so inexpensive that if California and Colorado quit suing each other over water and took the money devoted to legal fees and spent it on cloud seeding, they would have so much water that litigation would no longer be necessary. In answer to a "rain-stealing" question, Hazzard gave the same answer he gave in Washington: if they doubled the rainfall in an area, the extra amount represented just 1 percent of the moisture that was passing over the area.[79] Hazzard excelled at staying on message.

Not all of the attendees were impressed, least of all a USWB meteorologist, who downplayed Krick's World War II military service and attributed the silver iodide process to GE's Bernard Vonnegut and not to Krick. Meteorologists remained uncertain about precipitation processes, even if Krick did not. A proper meteorological survey of potential seeding areas would require special observations over several years. Krick was not doing that. And scientific controls over seeding? No chance that Krick's operation had them.[80]

Fed up, the meteorologist-in-charge (MIC) of the Idaho Falls Weather Bureau office laid it all out for a Federal Bureau of Investigation (FBI) agent. Hazzard's audience had included Idaho's governor, several congressmen, and some three hundred other people who had come to hear Krick. The MIC had "no doubt that Krick [was] operating a fraudulent organization for personal profit." Why? Krick was an "above average" meteorologist, but he had no answers when confronted with what was actually known about rainmaking. Therefore, Krick knew he was not on solid scientific ground and that his operations were dishonest. "Furthermore," the meteorologist wrote, "he has a long record of unscrupulous pseudoscientific activities." Based on Krick's use of inexpensive ground generators and the amount of land under contract, the MIC estimated that Krick's outfit would gross approximately $10 million ($100 million in 2015 dollars) during 1951. To make matters worse, Krick's group was getting observational assistance from the Pocatello, Idaho, Weather Bureau office by calling in at night to speak with the observer when the MIC was at home.[81] Could the FBI help?

By mid-July 1951, USWB chief Francis Reichelderfer was targeting Krick and his weather modification activities. In an administratively restricted letter to his security officer, Reichelderfer wanted to know if he should advise the US attorney general that Krick's nonprofit research group was operating too closely to his for-profit rainmaking outfit, which might make Krick guilty of fraud. The government, he wrote, was interested, and Krick was misleading thousands of farmers as he claimed to be the source of the water falling on their land.[82]

Following up on Reichelderfer's letter, the USWB security officer contacted the Idaho Falls MIC about Krick. How did Krick and other rainmakers induce agricultural interests and others to sign rainmaking contracts? How were Krick's research and operational organization integrated? Was the research group a nonprofit entity? If so, did it share in the profits from the operational side of the business? Furthermore, was the FBI investigating rainmakers in the area?[83]

Like gumshoes in a pulp mystery, the MIC and an assistant followed Krick's activities. They knew he had been working in their area because local farmers had been solicited through letters, brochures, and personal contacts. The farmers were then invited to joint meetings where they were encouraged to form groups that contracted with Krick's firm. While the company's representatives chatted up the farmers, its press office fed articles extolling the scientific legitimacy of cloud seeding to local media.

The plot thickened when the MIC attended one of these meetings led by an "ordinary farmer," who obviously was not. The "farmer" told the audience that he was ignorant of meteorology and could not explain processes for enhancing precipitation, but the company's representatives were standing by to study their particular situation and increase their rainfall. While the MIC's blood pressure skyrocketed, the man glorified Krick, cited his successes, and indicated that some of the farmers had been placed in contact with enthusiastic customers in New Mexico. The MIC was especially disturbed by the man's criticism of government agencies, the USWB in particular, in "vicious and particularly objectionable" language. The "ordinary farmer" implied that the USWB was attempting to corner the weather control market while simultaneously failing to develop new scientific methods for improving agriculture. This front man, the MIC believed, would be collecting a rather "fat commission" after the contracts were signed, and his "local accomplice" was important for the sales plan.

Like an evangelist issuing an altar call at a revival meeting, Krick's representative asked farmers to come forward with their checks to start the program. The farmers failed to show sufficient interest, so a local person offered to make rain. At that point, the MIC and his assistant stood and pointed out some of the falsehoods shared during the presentations. A closed meeting, which still failed to produce sufficient funds to engage Krick's firm, followed this outburst, and the contract went to the local rainmaker, who admitted that his technique did not always work. The MIC advised the USWB security officer that the Idaho Falls FBI office had taken no action against Krick. Officials of the FBI knew how Krick's organizations were operated, but they could do nothing to stop him. Since Krick had not gotten the contract, the Idaho Falls FBI office was no longer interested in investigating him and his firm, and it took no further action.[84] But Reichelderfer, who kept an extensive dossier on Krick, was still very much interested in learning more about him, and the MIC-turned-gumshoe continued monitoring Irving Krick's rainmakers.

By early 1952, Senator Clinton Anderson was claiming that rainmaking outfits were sending people into New Mexico to encourage farmers and ranchers to lobby against federal regulation of weather control. He noted that if people like Krick were not responsible for droughts and other unfortunate outcomes from cloud seeding that damaged crops and livelihoods, then they had nothing to fear from federal legislation. New Mexico governor Edwin Meachem, in office for just a year, decided not to ask his legislature to enact weather modification regulations because of state border issues. Since the operations and their effects extended beyond the state line, it was not within the state's purview to control them.[85]

While the situation in New Mexico remained quiet throughout 1952, by summer 1953 conflicts spawned by commercial rainmaking were beginning to heat up. Robert McKinney, editor of the *New Mexican*, was investigating rainmaking businesses operating in the state, and he found that commercial rainmakers were "closed-mouthed," and their burned customers "sheepish" and mostly unwilling to talk. After the charitable rainmaking bill (chapter 3) passed the New Mexico legislature, at least a dozen tax-exempt rainmaking organizations sprang up, most of them contracted with Krick, and for the next two years silver iodide smoke rose all over the state. The drought intensified, and ranchers sold their stock at prices approaching Great Depression lows. The last of the rainmaking contracts expired in August 1953, as President Eisenhower and a number of federal leaders visited New Mexico to view the latest dust bowl. By mid-October, New Mexico had received nothing from Washington, but nature had provided unassisted rain.[86]

Indeed, widespread drought conditions became a major incentive for developing weather control techniques. Colorado's Weather Control Commission recommended conducting a survey to determine if it should sponsor a rainmaking project, but no funds were available and the legislature was not likely to appropriate them. A Texas Republican national committeeman urged Colorado to join Texas by contributing to an interstate fund to investigate rainmaking, but Colorado governor Thornton, who argued that rainmaking funds should come from the agricultural interests and private utilities most likely to benefit, opposed it. With government support off the table, the commission suggested that agricultural organizations contribute to the fund.[87]

A year later, the *Washington Daily News* reported that eighty-three counties in hard-hit Colorado, New Mexico, Texas, and Wyoming had been declared disaster areas, and Eisenhower had been asked to declare Missouri, Georgia, Alabama, Oklahoma, Arkansas, Kansas, Kentucky, and Illinois as disaster states. The dry area extended east through the Carolinas and Virginia, which opened these states to the allure of weather modification. Although the drought was not as serious as that of the 1930s, the long-term forecast of above-normal temperatures and below-normal precipitation was worrisome.[88] Thirty-three more counties in the South were later added to the federal disaster area list.[89]

Had the early to mid-1950s seen normal moisture levels across the United States, it is doubtful that commercial weather modification would have taken hold as quickly as it did. However, agricultural interests without sufficient water will do whatever they must to save their crops, livestock, and livelihoods. As conditions worsened, states felt pressured to "do something" about the weather, and providers such as Irving Krick were standing by.

Each of these states took a different approach to handling uncontrolled commercial rainmaking. Let's return to the state of Washington to see how government entanglements with weather modification played out during the rest of the decade.

LEGISLATING WEATHER CONTROL: WASHINGTON STATE (1953-1957)

As drought-stricken states were sorting out their weather control woes, the state of Washington sought to rein in weather controllers by establishing a weather control board. Affected interest groups immediately chimed in with recommendations, but it was all for naught when the bill died.[90] Two years later, legislators took another stab at establishing a weather modification board and defining its duties.[91] Alas, this bill died as well. Another two years, another new legislature, another bill . . . third time's the charm, the bill passed and was signed into law. Weather modification and control were defined as the use of artificial methods to change, control, or attempt to change or control the development of clouds and precipitation in the troposphere (the part of the atmosphere that extends from Earth's surface to approximately eleven miles above the surface in the middle latitudes). The director of conservation and development would chair the new weather modification board, whose members, appointed by the governor, would include two faculty members—one each from the University of Washington and the State College of Washington (now Washington State University)—plus a person engaged in horticulture and a person engaged in other agricultural products. They would issue licenses and permits to would-be cloud seeders.[92]

Washington State's Weather Modification Board was not a perfect solution, but at least it provided a way to track weather control activities, answer citizens' and rainmaking firms' complaints, and incorporate weather control techniques into its water resources plans. Board membership gradually changed as additional groups requested a seat at the table, but a solid mechanism was in place. Meteorologists were able to inject scientific background into the deliberations, and they could access the data provided by the weather controllers as they attempted to determine what was happening in the atmosphere. For the other board members, the difficulties of providing preferred designer weather across the state were becoming clearer with each request and complaint.

We will leave Washington and its weather skirmishes behind for now. While the cloud seeders, power companies, agriculture interests, state residents, meteorologists, state officials, and politicians continued to wrestle with the weather during the 1950s, the meteorological community was trying to gain atmospheric knowledge that would allow it to determine whether weather control was possible or not. Would the meteorologists continue to play second fiddle to the lab-based Irving Langmuir? How would they dodge the charge that they were closed-minded? Next up: the meteorologists.

CHAPTER 5

The Meteorologists: Corralling the Research Agenda

It is reasonable to anticipate that further advances in our knowledge [of natural weather processes] will suggest other ways in which the relatively puny hand of man can modify the vast forces of nature for the common benefit.

HENRY HOUGHTON, MIT Meteorologist, 1954[1]

Federal and state politicians and agency leaders spent the 1950s focused on the national security implications of weather control—including the need to augment water supplies—as they wrangled over how, or even if, it needed to be regulated. Meteorologists, however, focused on understanding the underlying physics of precipitation processes. But the weather control juggernaut threatening to discredit their newly won scientific reputation tore them away from data and equations and deposited them into the science policy arena. In this relatively new role, some meteorologists found themselves developing policy statements for the American Meteorological Society, advising congressional leaders on weather control legislation, and serving on the Advisory Committee on Weather Control (ACWC), which would help to determine the future of state-controlled weather.

What did meteorologists know about precipitation processes in the early 1950s? Not much. They knew that about ten thousand average-size cloud droplets could fit on the head of a pin, and if 1 million of those droplets coalesced, the resulting drop would be heavy enough to fall out. They also knew that the Bergeron method (chapter 1) described the precipitation process in middle and upper latitudes, but not in the tropics. Meteorologists had also determined that dry ice particles spurred the coalescence process in supercooled clouds, but not in those with above-freezing temperatures, and that dry ice particles could successfully punch holes in supercooled stratus clouds. Silver iodide seeds, which also worked as artificial nuclei, could be sent aloft from ground-based generators if an inversion were not in place that prevented the seeds from reaching the clouds. (Under inversion conditions, air gets warmer with increasing altitude, and warm air at the surface only rises until it reaches equilibrium with the warm air aloft.) Meteorologists also knew that sometimes it rained even if ice crystals, natural or artificial, were not present. They did not know how ice crystals occurred naturally or how air moved within clouds. Consequently, they could not predict the trajectory of seeds, how mixing would affect their diffusion, how fast silver iodide could work before sunlight reduced its effectiveness, or even how many freezing nuclei were already present in the atmosphere. Because of these uncertainties, meteorologists could not use physical reasoning to determine whether seeding produced precipitation or not.[2]

To understand precipitation processes, meteorologists needed to do much more research *in the atmosphere*. Langmuir kept arguing that laboratory work was key to solving the weather control puzzle, but meteorologists knew that only in-situ experimentation would yield needed data. In the meantime, their professional community needed to influence the national discussion about weather control.[3]

DEVELOPING POLICY: THE AMERICAN METEOROLOGICAL SOCIETY

As the weather control controversy intensified, the American Meteorological Society (AMS) appointed a committee to investigate related claims with the goal of creating a policy statement. The committee members represented three sectors of the meteorology community: Chairman Henry G. Houghton from MIT represented academe; Henry T. Harrison, manager of weather services for United Air Lines, represented commercial interests; and Sverre Petterssen, of the Air Weather Service, brought insights from the military/federal government.[4] Although most meteorologists downplayed weather control's possibilities due to their slim understanding of precipitation mechanisms, meteorologists had not formed a monolithic opposition group. Some were cautiously optimistic, some were providing weather control services, and some who opposed it were just as worried about the discipline's professional reputation as they were about weather control's validity.

The professional implications burst into the open during an AMS council meeting when a member attacked Henry Harrison's more favorable disposition toward weather control, charging that it would eliminate "all scientific progress made during the past twenty-five years." Meteorologists were painfully aware that their discipline had not been considered a "real" science until after World War II when the scientific community and the public acknowledged weather forecasting's importance for military success. Some AMS members thought venturing into a scientifically dubious undertaking like weather control could take the discipline down a few pegs to the "guessing science" category of the nineteenth and early twentieth centuries. Weather control's tight connection to the military with its concomitant secrecy also threatened to impede scientific progress in the atmospheric sciences, especially if military security continued to restrict the release of experimental data—a concern many US scientists shared during the Cold War.[5]

Meteorologists' primary worry, however, focused on the lack of scientific evidence supporting weather control efforts. As an airline employee, Harrison recognized that effective weather control techniques could aid aviation. He urged his colleagues to avoid the "negative, reactionary, and unrealistic" positions that many people thought the meteorology community held while they assessed the evidence. Harrison was concerned that the USWB and Air Force's distinct lack of curiosity during the Project Cirrus experiments would not serve the discipline well.

Harrison also brought news from the West. Undeterred by weather control's shaky scientific underpinnings, residents of semi-arid regions were contemplating how they might exploit a new economy based on artificial nucleation. Needing water, westerners were not keen on the government restricting weather control. Government officials who thought those living downwind from seeding areas were being harmed should prove it. "Weather associations" should be able to work out any problems occurring along state lines, and the federal government should butt out. Most press attention had focused on large rainmaking operations, but quieter work was underway. Actuarial associations were interested in hail suppression that might reduce insurance payouts, forestry departments were considering lightning suppression to reduce fires, and some airports were pursuing fog dispersal. To counteract the Boston-based AMS's eastern US focus, the Denver-based Harrison's mission was to bring information about weather control from the West and encourage a more sympathetic approach.[6]

But after Harrison and Houghton testified before Senator Clinton Anderson's joint subcommittee in March 1951 (chapter 3), Harrison rued the futility of their efforts. They had taken a middle path on weather control, but Anderson had put them "far to the right and *just one notch to the left*" of the US Weather Bureau's extremely dismissive view. He fretted that the USWB had been "discredited and humiliated" at both the formal hearings and an informal rump session Anderson had held with USWB personnel and a large number of audience members. Harrison was especially concerned that Anderson did not understand that cloud seeding could extend beyond state lines, but not one thousand to two thousand miles downstream.[7] Houghton went further, convinced Anderson and other committee members had thrown their weight behind weather control in advance of the hearings. No fan of weather control, he was starting to "waver." Scientists needed to stick to scientific facts, but should not reject the future possibility of large-scale weather control.[8]

Consequently, Houghton toned down the strident language in his initial draft of the AMS weather control statement. He changed "seeding will not draw water from a stone" to precipitation required pre-existing moist air and appropriate cloud-forming processes. Not everyone reading the statement would be meteorologically savvy, so he defined "large scale" as most of the country, not a few counties. Houghton emphasized the need for cloud physics experiments designed by scientists possessing expert knowledge of cloud physics, synoptic meteorology, and statistics. They had "missed the boat" on the Senate hearings and needed to be ready for the upcoming House hearings.[9] Houghton, Harrison, and Petterssen, recommending that a committee of "eminent" scientists be appointed to thoroughly evaluate rainmakers' claims, supplied copies of their report on weather control to Anderson and other relevant lawmakers.[10]

Their middle-of-the-road approach would not please everyone. Rainmakers would dub it "reactionary," while AMS's conservative factions would find it "radical." Harrison thought the draft section on licensing requirements and qualifications was too severe: "sincere and true pioneers" in cloud seeding who lacked scientific training but approached it in a scientific way would be shut down. A prime example: Weather Control, Inc., in Medford, Oregon, which had a fog-clearing contract with United Airlines. How could these men continue their valuable work while AMS eliminated "rank amateurs and quacks"? Even Langmuir and Schaefer would have been kept from their experiments since neither was an AMS "professional meteorologist."[11] Houghton conceded the point, but AMS needed to keep the quacks out. If weather control became big business, "it [would be] unlikely that many qualified meteorologists [would] have the capital to enter [the field]." AMS could not require individual and corporate seeders to become members, but it could suggest that they meet the same requirements as AMS's professional members.[12]

Houghton et al. reported to AMS secretary Charles F. Brooks on the problems inherent in weather control legislation. They did not want to be perceived as a "minority pressure group" or as dictating legislative language; they just wanted to encourage congressional leaders to base legislation on scientific principles and public protection. AMS Council members might not think weather control legislation was inevitable, but Houghton and his colleagues were convinced it was coming. It was too late to influence Senate bills, but they still had a chance in the House. Recent floods, which some thought had been triggered by seeding, had given Congress additional impetus to pass legislation.[13] The AMS's Statement on the Legislative Aspects of Weather Modification favored federal legislation that matched the "small scale" status of current weather modification. It also recommended federal licensing of rainmakers, perhaps using AMS's professional guidelines. The federal government should carry out expensive cloud physics experiments, which would be more likely to produce scientific knowledge than data produced by commercial rainmakers. AMS argued that additional research would allow scientists to determine if controlling the weather were possible.[14]

This statement addressed only legislative approaches, so Houghton et al. continued their work on an AMS policy statement. True to his word, Harrison continued to forward information on events in "weather control country." Significant seeding programs were underway in major watersheds—Colorado, Arkansas, Rio Grande, South Platte, and North Platte—and contracts covering eastern Colorado had been renewed. Harrison thought seeders might be "lying low" because of liability. Major snowfalls during the 1951–52 winter had closed highways, isolated some towns for weeks, and resulted in loss of life. The synoptic situation had been ideal for heavy snows, but Harrison thought some people would claim seeding had triggered them.[15]

Harrison was keeping an eye on a weather control evaluation bankrolled by a $100,000 ($1 million in 2015) donation from "rain-enhancer" Irving Krick and conducted by UCLA meteorologist Joseph Kaplan. Kaplan had an excellent reputation, but when the National Weather Improvement Association (NWIA) sponsored a roundtable discussion on his findings, Harrison was discomfited that the "objective and independent" NWIA was composed entirely of Krick's rainmaking clients.[16] Houghton et al. were eager to read others' evaluations as long as weather modification consultants were not influencing them.

By fall 1952, Petterssen was rethinking that position. Based on a report written by meteorologist Paul MacCready, who had been conducting weather modification tests in Arizona, it might be possible to increase rain over a large county or a small state. It might be correct to say that the evidence was inconclusive or showed a few positive effects. Petterssen remained skeptical of Langmuir's claims of large-scale effects due to periodic seeding, but could not discount them completely. He still wanted to stay clear of licensing and regulations. He wondered if cloud seeders were "quacks or scientists." "Do we want to license quacks?" And even though he favored the latest bill (S. 2225), he still opposed military involvement in any legislation. According to Petterssen, the military did not need permission to conduct experiments that would benefit the nation. Any temporary committee assigned to evaluate weather control would also need a longer life span—at least five years, not the two proposed by congressional sponsors.[17]

As they polished the final draft, Harrison worried about how they would frame the statement to avoid the "bad press" received by the USWB. He wondered if it were possible to change the climate in a small area. Certain topographical areas might be ideal for seeding: mountainous regions along coastlines where upslope winds occurred regularly, or peninsular areas where convergent wind flow would provide clouds.[18] Harrison was trying to narrow down the circumstances under which they might give a positive nod to weather control while not indicating that it was effective over large swaths of the country.

With the final draft imminent in early 1953, AMS secretary Brooks began polling council members to see which way the wind was blowing on weather control. The answer: predominantly light and variable. Some supported the draft statement, but admitted to knowing little about weather control. Some were undecided and nitpicked scientific details. The consistently disdainful USWB chief, Francis Reichelderfer, was generally supportive, but thought it might be interpreted too positively. Air Force brigadier general Benjamin G. Holzman declined to vote, but agreed with a *Washington Post* headline: "Rainmaker says carbon dioxide not needed—only water required for rain." Others wanted much more meteorological detail about the atmospheric prerequisites for artificially inducing precipitation.[19] In other words, councilors' opinions ranged from indifference to requirements that would have made the statement unintelligible to nonmeteorologists.

Houghton, Harrison, and Petterssen dutifully included suggestions in the final revision, but remained uncertain of its value. Congress appeared ready to pass a weather control bill that would create a commission to study cloud seeding and make legislative recommendations. AMS needed to be proactive and develop a list of potential commission members. The society needed to be seen as having a "liberal, progressive viewpoint," or government officials grappling with weather control would not pay it the least bit of attention. Meteorologists were the subject matter experts, but their "reactionary and negative" comments had been ignored. The AMS policy statement needed to overcome the perception that meteorologists held a negative attitude even as they included relevant scientific facts. The "peculiar coincidence" of the weather periodicities that Langmuir continued to trumpet to anyone who would listen required leaving the door to weather control "slightly ajar" instead of slamming it shut with a bang.[20]

After the AMS released its Statement on Weather Modification and Control in May 1953, members, including at least one who had voted affirmatively, started throwing darts at it. Some disputed the assertion that silver iodide seeding had not yielded statistically significant increases in rainfall, referring to an obscure publication known only to those who worked in the West.[21] One member thought the statement lacked the "quality of scientific open-mindedness."[22] Houghton took these and other comments under advisement, but declined to make any changes. If AMS watered down the wishy-washy statement even further, it would be too weak to be of value. A statement that the entire membership agreed with would be meaningless.[23] While the AMS wrestled with its official position on weather control in the face of continued uncertainty—does it work or not—a number of research groups within and outside the United States were trying to put the science of weather control on firmer ground.

RESEARCH ON WEATHER CONTROL

Within the United States, weather control research took several paths during the 1950s. Some meteorologists attempted to develop viable theories on cloud physics and precipitation mechanisms, while others concentrated on the practical aspects of creating designer weather. Funding came from the federal government, foundations, and university-related research centers. Statisticians played a more important role assessing the statistical significance of precipitation increases from seeding. Outside the United States, substantial projects were underway in Australia and Israel, which needed fresh water. Weather control research and efforts in "friendly" nations were not viewed as a problem, but Soviet research was seen as a direct threat. Therefore, US researchers capitalized on Soviet efforts, not because the American meteorologists knew what the Russian meteorologists were doing, but because they did not. That uncertainty kept the funding spigots open for weather control projects that could be transformed into offensive and defensive weapons during the Cold War.[24]

Projects in the United States

By late 1951, the Research and Development Board's Special Committee on Cloud Physics concluded that the basic principles underlying weather modification techniques were sound; some small-scale processes could be changed artificially. However, its members did not agree on the extent to which weather processes could be changed by seeding with pulverized dry ice or silver iodide seeds. The military services wanted to use weather control to dissipate fog, open holes in stratus clouds, induce precipitation from cumulus clouds, increase rainfall over large areas, and steer hurricanes. They had no operational techniques for the first two, hail mitigation (related to seeding cumulus clouds) continued to be inconclusive, and widespread seeding had yielded mixed results; in 1947, a seeded hurricane had changed direction, but proof of the cause was lacking. The purported periodicity in rainfall claimed by Langmuir during Project Cirrus could not be proved nor disproved. If periodic seeding did influence continental-scale weather, then it should be possible to modify large-scale circulation patterns.[25] Possibilities? Yes. Operational techniques? No.

Therefore, committee members recommended (once again) additional laboratory research to explore specific properties of artificial nuclei and field research to explore cloud droplet growth and coalescence. Weather control R&D needed to be commensurate with its potential military and economic value. Dissipating supercooled stratus clouds or fog with the goal of creating an operational technique should be an immediate focus. Tactical rainfall intensification was more problematic because of cloud variability. Enhancing rainfall for agriculture made more sense because it could be beneficial despite a lack of pinpoint accuracy. Seeding frontal systems as they moved across the country and into "seed-free" zones could help researchers get a better sense of seeding's effects.[26] The Artificial Cloud Nucleation (ACN) projects, which encompassed all of the military services and the USWB, emerged from this final suggestion.

In aggregate, the ACN projects were SECRET—even though the USWB's section was unclassified, the Army and Air Force projects were RESTRICTED, and the Navy's was CONFIDENTIAL. Why? If viewed together, one might see a state-funded, broad-scale attack on a weather control problem of military interest.

The separate Office of Naval Research–funded "Project Shower" involved a multidisciplinary team that coordinated its efforts with the Pineapple Research Institute, the Hawaiian Sugar Planters Association, and the Radiophysics Division of the Commonwealth Scientific and Industrial Research Organization (CSIRO) of Australia. These federally funded projects are listed in table 5.1.

Research universities in the United States were also carrying out weather control investigations. The University of Arizona's Institute of Atmospheric

TABLE 5.1. Weather Control Research in the United States (federally funded projects, 1950s)

Name	Directed By	Experimental Goals
Artificial Cloud Nucleation (ACN)	Army Signal Corps/ Bernard Vonnegut	• dissipate warm fog and low-level stratus • modify lattice structure of silver iodide crystals to match ice crystals • study effect of sunlight and atmospheric contaminants on silver iodide seeds • alter seeds to make them effective at higher temperatures
	Army Signal Corps/ Helmut Weickmann	Examine problems related to: • overseeding • lateral spreading of artificially induced crystals • the effects of double seeding • the effects of different reagents (silver iodide, Freon, compressed air)
	Air Force/University of Chicago	• study physical and chemical attributes of warm cloud precipitation • conduct field studies and develop instruments to study natural precipitation
	Weather Bureau	• study natural precipitation mechanisms and effects of cloud seeding on cloud structure and precipitation related to Pacific Coast storms in Washington State
Project SCUD	Navy	• detect seeding effects on the formation and deepening of storms near the Eastern Seaboard
Project Shower	Office of Naval Research	• examine precipitation mechanisms in Hawaii, including the role of sea salt in rain formation

Physics and the University of Chicago's Meteorology Department were two major academic institutions conducting cloud physics and artificial nucleation research. The Tucson area intrigued atmospheric scientists because radar images showed rain was initiated high in the clouds at low temperatures, whereas in the eastern United States, half of the radar echoes indicated rain was initiated at above-freezing temperatures. The Desert Southwest was usually in need of rain, and Tucson provided the ideal place to determine how nature made it. If it were possible to improve the rainmaking mechanism, it could mean a significant economic payoff for semi-arid regions worldwide. Tucson was also ideal because it was often the site of easier-to-test isolated

TABLE 5.2. Weather Control Research in the United States (1950s): University-centered Research

University	Experimental Goals
University of Arizona, Institute of Atmospheric Physics and University of Chicago Meteorology Department	• conduct a cloud census • determine natural precipitation mechanisms, including level of initial precipitation in clouds • count freezing nuclei in clouds • perform a synoptic climatology study
University of California, Berkeley, Statistics	• test hypothesis: cloud seeding does not increase precipitation • examine medium-scale (county/small watershed) seeding results • determine method of randomizing experiments

cloud systems. Statisticians from the University of California, Berkeley, were also examining weather control. Unconvinced of cloud seeding's efficacy, they sought to create testing methodologies that would be statistically sound. (For details, see table 5.2.)

Research Abroad

Cloud physics and artificial nucleation experiments were underway on every continent save Antarctica, as national meteorological services attempted to tap atmospheric moisture. In some nations, government meteorologists conducted the research, and in others, commercial rainmakers from the United States did so. American meteorologists involved in cloud physics research readily shared information with their foreign colleagues, even if their colleagues involved in the classified realms of weather control could not. (See table 5.3.)

However, weather control efforts underway in the Soviet Union were the real drivers of interest in weather control in the United States—and the reason federal funding was available for their work. United States meteorologists had long speculated that their Soviet counterparts were ahead in weather control efforts, but they had limited or no access to Russian-language publications on weather control research until the late 1950s when translations became available. (Earlier translation efforts had given priority to nuclear physics.)[27] Based on the works cited in their publications, Soviet scientists had access to non-Russian language publications, probably much better access than western scholars had to theirs.[28]

TABLE 5.3. Weather Control Research Outside of United States (1950s)

Continent/Nation	Directed By	Experimental Goals/Results
Africa		
Tanganyika (now Tanzania)	East African Meteorological Department	Goal: induce rain by seeding clouds with thumb-sized "bombs" filled with gunpowder soaked in silver iodide Result: overseeded; more rain on non-seeding days
Europe		
France	Observatory of Puy de Dome	Study: • effect of sunlight on silver iodide strength
Italy	Aeronautical Meteorological Observatory, Mt. Cimone	• detect presence of silver iodide particles in the free atmosphere
Sweden	University of Stockholm, Meteorology Department	• enhance rainfall by seeding with crop-dusters in the mountains
Switzerland	Polytechnic Institute, Zurich	Study: • crystal structure of silver iodide • modification of silver iodide molecules and their effect on ice nucleation and deactivation • the effect of rocket propellants on ice nucleation • hail prevention
United Kingdom	Imperial College	Study: • relationship of nuclei to cloud particles and subsequent development of rain and snow
Middle East		
Israel	Ministry of Agriculture/ Meteorological Service/ Israel Air Force/Weizmann Institute of Science	Study: explore possibility of producing artificial rain in Israel Results: Mixed

(*continued*)

TABLE 5.3. (continued)

Continent/Nation	Directed By	Experimental Goals/Results
		South America
Peru	US-Owned Mining Company/ Wallace Howell	Goal: increase runoff for hydroelectric power by sprinkling coal dust on snow Results: three times as much runoff from blackened snow Goal: spray clouds with silver iodide solution to induce showers Result: induced showers contributed to runoff for irrigation at a distant plantation
		South Asia
Pakistan		Goal: introduce salt particles to clouds by spraying a salt solution on a dirt road near the Khyber Pass Result: up to two inches of rain after spraying
		East Asia
Japan	Central Meteorological Observatory/universities/power companies	Goal: induce rain for hydroelectric power Study: fundamental rainfall mechanisms
		Australia
Australia	Commonwealth Scientific and Industrial Research Organization (CSIRO)/Edward G. "Taffy" Bowen	Goal: determine if meteors were showering the Earth with cosmic dust that served the same purpose as artificial seeding
		Eurasia
Soviet Union		Goals: • disperse clouds and fog, particularly in the Arctic, to aid aviation and ship traffic • prevent hail, catastrophic floods, and eroding downpours from convective clouds

At this same time, US newspapers started picking up reports about Soviet uses of weather control and then hyping the possible danger of the Soviet Union gaining control of the weather and indeed the world. A United Press International dispatch in late 1957 reported that the Soviets had produced rain clouds in a laboratory by using radioactive elements, an interesting combination of the weather and nuclear threat rolled into one. Referring to an article in the Soviet newspaper *Zaraya Vostoka*, the dispatch indicated that the scientists had used a "complex electrical and radiochemical process" that produced tiny water particles that eventually grew into a cloud that rained . . . all in a specially built chamber.[29] Not only were the Soviets going to make rain on demand, they were going to make radioactive rain. Soon thereafter, USWB chief Reichelderfer expressed his concerns about the Soviets' growing interest in weather control and their expansion of basic meteorological research.[30] And a page 1 *New York Times* article based on extensive interviews with military personnel suggested that to preserve American military superiority, projects focusing on the feasibility of weather control "must receive strong support."[31] Despite an apparent lack of intelligence on Soviet weather control, US officials were convinced that the Soviet Union was ready to dominate the world with its extensive weather control techniques while at the same time Americans downplayed the Soviet system's ability to produce good science.

Soviet meteorologists had been conducting weather control experiments since before World War II, but apparently they were all small-scale. They chose not to distribute their results widely—or more likely, the Communist Party and the Soviet government decided they would not publicize their results outside the Soviet bloc—and West bloc countries interpreted that as "being ahead." The Soviets were conducting cloud physics experiments as they gathered basic information about precipitation processes, but they focused more on clearing clouds and fog than on inducing or enhancing precipitation.

The extent of state support, encouragement, or coercion involved in weather control efforts around the world depended on the type of government and its needs and wants. Most countries were interested in bringing more water to dry lands, but some, like the Soviet Union, were more concerned with keeping their crops from being beaten down by hail. While the Soviet Union kept its meteorologists on a short leash, other nations took a more relaxed approach to weather control efforts. In the United States, with commercial firms taking the lead on the domestic front and military units taking the lead on possible uses of weather control outside the nation's borders, by the mid-1950s it was time to evaluate the situation and develop a national plan of attack. The Advisory Committee on Weather Control was established to do just that.

THE ADVISORY COMMITTEE ON WEATHER CONTROL
(1953-1957)

The ACWC—essentially a consolation prize for Senator Anderson when he could not get his Weather Control Commission bill approved—was to survey the weather control situation, evaluate extant data, and provide recommendations to the president.[32] By early September 1953, the White House was seeking nominations of appropriate persons to fill the positions reserved for private citizens, not only from relevant cabinet secretaries and the National Science Foundation director, but also from the chairman of the Republican National Committee.[33] A number of government officials and all of the cabinet secretaries provided suggestions; many of the nominees were already serving on various weather control-related committees.[34] The White House sent the compiled list containing the names of twenty-two distinguished men from a variety of fields to the relevant cabinet secretaries for comment.[35] The final selectees were the chairman, retired Navy captain Howard T. "Shorty" Orville, a meteorologist and consultant for Bendix Corporation; Kenneth C. Spengler, executive director of the American Meteorological Society; A. M. Eberle, dean of agriculture at the South Dakota A&M (now, South Dakota State University) and member of South Dakota's Weather Control Committee; former congressman, director of the Bureau of the Budget, and ambassador to the United Kingdom Lewis W. Douglas of Arizona, a prominent rancher and banker; and Joseph J. George, superintendent of meteorology for Eastern Air Lines. Once they were chosen, the White House forwarded their names to the Republican National Committee as a courtesy. Only two of the selectees were Republicans: Eberle and Orville. Spengler was an independent, Douglas and George were Democrats, but all three had supported Eisenhower during the 1952 election.[36] President Eisenhower formally appointed them in December 1953, and the Senate confirmed them in January 1954.[37]

Not waiting for funds and a staff, Orville and his committee quickly defined their primary tasks: gathering data from both public and private cloud seeders and evaluating the results; investigating the extant knowledge of atmospheric physics and the probable effects that people could have on atmospheric phenomena; examining the legal and economic aspects of artificial manipulation of atmospheric phenomena or patterns, and developing a recommendation regarding appropriate federal regulations; and considering the most viable plan to pursue both basic and applied research in atmospheric physics in the public interest. It would undertake these tasks with the assistance of "outstandingly able" meteorologists, physicists, statisticians, economists, and attorneys who would serve as consultants or as committee staff members.[38]

By early June 1954, committee members had been thoroughly briefed by the Department of Defense (DoD) and had held three additional meetings during field trips to weather modification project sites: Ramey Field, Puerto Rico, where the University of Chicago (under contract to the US Air Force) was conducting randomized seeding experiments on warm cumulus clouds; Seattle, Washington, to visit the University of Washington's Meteorology Department, discuss items of state and local interest related to weather control, and check out the USWB's ongoing Artificial Cloud Nucleation project; and Salt Lake City, Utah, to attend the Western Snow Conference and to discuss, with University of Utah researchers, rainmaking experiments underway in southern Utah, started at the behest of ranchers who had engaged Irving Krick's Water Resources Development Corporation to do the seeding.[39]

The ACWC also queried government agencies about their interests in weather control (chapter 3). Water supply consistently led their lists. The US Geological Survey's (USGS) Luna Leopold, a distinguished geomorphologist and hydrologist, argued on behalf of the Interior Department that it was extremely important to neither over- nor underestimate weather control's potential. Objective, critical analysis and evaluation of weather control techniques were of the utmost importance. Interior did not want overly optimistic statements in the press that would encourage people to bombard them with requests to alter existing water resources plans, or to open up marginal lands for cultivation based on this "new-found" water supply. For this reason, Leopold argued that Interior might be more concerned about weather control's viability than other departments.[40] Therein lay the committee's problem: how could it sift through contradictory results from government and commercial weather control studies and make recommendations that would not potentially lead to widespread harm in the years ahead?

The National Science Foundation was the one government entity standing by to support basic research in weather control, and it argued that many of the ongoing technological efforts were probably premature given the lack of fundamental knowledge behind nucleation, the formation and dissipation of clouds, and atmospheric physics. The NSF recommended more laboratory and field research, and it suggested putting the engineering portion of weather control on a back burner until scientists had a better understanding of basic principles and had conducted "carefully planned field experiments." Willing to host conferences on basic atmospheric science and weather control, the

NSF was ready to assist other government agencies that wished to modify the atmosphere for their own specific purposes.[41]

After five months, the ACWC was already trying to make sense of what it had learned. Although DoD was not the only executive department interested in weather control, it was singularly responsible for all federal weather control activities, which at the time were devoted to artificial cloud nucleation. However, none of the Artificial Cloud Nucleation projects were focused on arid or semi-arid regions, which the committee thought needed immediate attention. As USWB chief Reichelderfer pointed out, projects in low-moisture areas most closely approached laboratory conditions. The USWB's part of the project in Washington State was being "complicated by the presence of too many nuclei" in the clouds. However, other committee members fussed over the implication of telling DoD how to carry out its research projects and use its assets. Perhaps they could recommend that other interested departments step up and undertake weather control research directly related to their needs, but those departments would still have to rely on military resources. Who besides the Air Force had air crews and specially outfitted aircraft to support weather control experiments?

Although USGS's Leopold thought it was premature to recommend an operational program to combat drought, members did note Senator Case's recent comment that compared appropriating funds for drought relief ($15 million) with the ACWC's budget request ($200,000). The former, he said, was like "paying for a dead horse." While the amount of money devoted to weather control research was significantly smaller, it could yield big dividends down the line if weather control could prevent or ameliorate drought conditions. Some committee members thought they could use the drought expenditures, which would surely top the most recent $15 million appropriation, as part of their argument for more federal investment in weather control research.[42]

In the early months of their investigation, committee members had noted two serious problems: no controlled weather modification programs were underway in arid regions of the country, and DoD intended to withdraw aircraft and air crews from ongoing cloud physics research programs by late November 1954. The latter was especially worrisome because DoD was in the best position to provide these resources, but the Eisenhower administration wanted to keep government costs down.

The committee had discovered that DoD was the primary government patron for cloud physics research and artificial cloud nucleation experimentation. It conducted some of the work in its own laboratories and either contracted out the remainder to universities and private companies or transferred funds and/or aircraft to the USWB so its meteorologists could carry out research projects. Although it was not within their portfolio to monitor research activities or to make recommendations related to research policy until their work was complete, the ACWC members were conscious of their responsibility to evaluate weather modification experiments. During this process they had uncovered some "uncertain and controversial scientific aspects" of weather control, concluding that the uncertain bits could be resolved with more research and experimentation on theoretical cloud physics and applied weather modification.

Therefore, the ACWC members wrote directly to President Eisenhower expressing their desire for continued federal support for ongoing and future research efforts. Committee members were particularly vexed by drought conditions in the nation's arid lands and argued that undertaking artificial nucleation research in these areas could simultaneously produce basic scientific knowledge and provide significant economic benefit to residents. Furthermore, they found it "inconceivable" that the results of such research would not have appropriate military applications. Private monies had been devoted to some of these research projects, but they were insufficient to cover aircraft and crews. The committee recommended taking immediate steps to "preserve an effective Federal weather control program" by letting DoD continue its cloud physics research programs, developing a weather modification project that would aid drought-stricken areas, and providing air crews and aircraft to the Artificial Cloud Nucleation programs and similar government-sponsored weather control activities.[43] The committee members did not intend to meet with Eisenhower to discuss their recommendations, but Orville, Lewis Douglas, Dr. Alan Waterman (NSF), and Donald A. Quarles (assistant secretary of defense for research and engineering) did sit down with the president for a fifteen-minute meeting that stretched into thirty-five minutes due to Eisenhower's interest in weather control. During the meeting, Douglas stressed the need for military aircraft support for the new program focused on arid regions being developed at the University of Arizona. Not only did Eisenhower ask Quarles to "take another look" at the possibility of meeting the ACWC's request; he also ensured that Quarles would convey Eisenhower's personal interest in the program.[44] The University of Arizona got the needed aviation support.

Over the next eighteen months, the ACWC members received input from several consultants who were examining the nation's short- and long-term water needs, drought suppression, and scientists' comments and recommendations.[45] They also considered studies that discussed psychological attitudes toward and religious disapproval of weather control, which were not likely to be dispelled by scientific proof and should be "regarded as unlikely to be completely overcome." Possible legal entanglements between groups trying to produce different types of weather over the same patch of land were another concern since it was unclear whether ownership of water resulting from seeding would "refer to airborne moisture or to that deposited on the earth." Riparian doctrine—which stipulated that each owner of land contiguous to a stream may use it for irrigation, watering livestock, and domestic purposes—was not going to work because clouds were not confined to a channel, and one would not be able to seed over land that would not benefit from seeding.[46] While the ACWC members tended to focus on weather control's technical aspects, societal and legal issues, which were becoming increasingly complicated, needed their attention as well.

Throughout 1955, evaluations of ongoing weather control experiments in the United States, Hawaii, and Puerto Rico poured in. In addition, an ACWC panel led by meteorologist Eugene Bollay analyzed possible effects of atomic and thermonuclear explosions on the weather, a topic of some concern among the general population, but less so among most meteorologists, who failed to find a direct connection between the two. John von Neumann suggested that debris from the explosions might be lifted into the upper atmosphere, much like ash from volcanic eruptions, thereby leading to temporary global cooling. However, Bollay seemed to think that nuclear blasts might lead to more direct effects downwind from the blast site as particles remained in the lower atmosphere where they might induce or prevent precipitation depending on the atmospheric situation. He suggested that instead of taking an "irreversible negative stand" on large-scale weather effects due to seeding or nuclear blasts, it would be preferable to engage in a five- to ten-year assessment of small- to large-scale weather control options, especially considering their possible economic impacts.[47] The ACWC also directed an effort to determine if seeds from ground-based silver iodide generators would be carried into supercooled clouds by natural air currents and produce the desired results: enhanced rainfall or, in the case of preventing rainfall, drying out the clouds. "Project Overseed" was carried out at the Mount Washington Observatory in New Hampshire, and preliminary results were ready in late 1955.[48]

Originally, the ACWC was to conduct its review of weather control in two years, which many considered too few. Nevertheless, the committee had to provide a final report on its work by June 30, 1956. Hoping for an extension, the committee members filed an interim report in February 1956. In the cover letter to President Eisenhower, they pointed out that under their supervision a superior methodology for statistically evaluating weather control experiments had been put into place; an extensive series of evaluations had shown that cloud seeding produced substantial and economically important increases in precipitation in the Pacific Coast states; and a physical evaluation program had produced valuable results and given assurance of providing additional information that would be vital to solving the weather control problem. Committee members also argued that the United States was facing a "pressing water problem," which, as Eisenhower had stated repeatedly, would "grow more desperate in future years." They assured him that the ACWC could be a "factor in solving the nation's water problems."[49]

The committee reported progress on its four primary tasks, but it still wanted to see observable results from controlled experiments in addition to data coming in from commercial seeders' reports. The ACWC members were reluctant to adopt conclusions on silver iodide's effectiveness without significantly more experimental data. They had already concluded that the federal government needed to take a major role in supporting academic manpower and basic research to advance knowledge of atmospheric processes. Economic and legal aspects of weather control were difficult to assess while atmospheric understanding remained weak, but committee members were not convinced that federal regulations were needed. While encouraging basic and applied research was outside the committee's purview, it held that the National Science Foundation should support studies of natural and artificial nuclei, cloud droplet kinetics and electrical effects, and thermo- and hydrodynamics of the lower atmosphere. Various cabinet departments were already supporting applied research related to their activities.

The committee also recommended that its original two-year mandate be extended for another two years, not because it had "failed to get the job done," but because it had "succeeded in establishing some positive and important results which justify the Federal Government continuing its special interest in the field." Within the additional two years, it would continue statistical evaluations of seeding projects; seek to find common meteorological factors in commercial operations that yielded outstanding statistical results by themselves or during certain types of storms; stimulate and encourage basic research into precipitation processes; seek reliable information on practical methods of suppressing lightning and hail; continue studies of commercial seeding along the Pacific Coast to determine orographic effects on precipitation; and, lastly, determine the frequency of favorable seeding situations within defined regions.[50] The committee argued that much more weather control research was required because of its potential importance to every citizen and to the nation as a whole.

As one might have predicted, USWB chief Reichelderfer was not particularly happy with the interim report. Writing to NSF's Waterman, Reichelderfer held that the committee had "exaggerated the findings" when it reported a significant "breakthrough" in cloud seeding. He wondered how he should advise the under secretary of commerce, a committee member, who might not want to be party to this announcement unless he knew of statistical evidence that was much clearer than had been previously reported. "The impetus given by these statements to wild-cat commercial seeding," Reichelderfer wrote, "and the credence at home and abroad to a report associated with the White House are matters of considerable concern."[51] In a letter to ACWC member Lewis Douglas, Reichelderfer wrote that the report gave the positive side of weather control without "giving the public the benefit of the cases that showed negative results." Since the press conference that accompanied the report's release, the USWB had been peppered with queries concerning rainmaking. It had been making measured responses, but it appeared to the public that the USWB's silence meant it "subscribes to the publicity." That the ACWC had found significant increases only in the mountainous areas of Pacific Coast states had been lost in the announcement. "Shall a committee at the Presidential level," Reichelderfer asked, "lead the public to believe that a solution to rainfall deficiencies has been found when in fact the case is still full of uncertainties?" Was the ACWC an "evaluation committee or a promotional agency"? "The public," Reichelderfer maintained, had "a right to know whether it [was] reading a sales talk or carefully weighted facts."[52] But as the committee's Charles Gardner noted, members had known before the report's release that their "conservative friends" would react negatively if they put out any positive information, and on the whole thought the report and publicity had been well received.[53]

Extension!

Even before the interim report's release, Senator Warren Magnuson (D-Washington) introduced a bill extending the ACWC for two years. The only cabinet department not supporting the extension: Commerce, home of the USWB. Commerce secretary Sinclair Weeks suggested that the USWB, DoD, and other agencies and universities had run sufficient tests and secured sufficient data to move ahead with basic research. He saw no reason to continue the ACWC, an opinion likely heavily influenced by Reichelderfer. In contrast, Interior was so worried about the nation's water supply that it argued for the extension to allow the ACWC to complete its evaluation and thus provide valuable information leading to better water resources, as well as land-use planning and programs.[54] Magnuson's bill passed in July 1956, and the new reporting deadline was June 30, 1958.

The committee continued its work during the six months that the extension was in play. This period's major event: the Conference on the Scientific Basis of Weather Modification Studies, held at the University of Arizona's Institute for Atmospheric Physics. The thirty-five participants comprised an international who's who of scientists and statisticians who had been involved in weather modification since the mid-1940s. They shared a series of still-unanswered questions about weather modification and cloud physics, with the ACWC's positive evaluation of statistical data pertaining to increased precipitation on the Pacific Coast a major indication of "significant positive results."[55] However, some statisticians challenged these results, taking aim at the lack of randomization in the experiments, how storms were selected for seeding, and possible errors in the analytical statistical models. Their charges were promptly rebutted by those involved with the ACWC's own statistical program, a fracas that migrated from the conference into the local, and then national, press.[56] As assessed by University of Arizona geophysicist James E. McDonald, the conference papers and discussions yielded less progress in cloud seeding than had been touted, but revealed considerable interest and labor going into cloud physics and precipitation studies. Extant cloud-seeding techniques suffered in three distinct ways: increased rainfall was not easily observed; seeding effects and potential may not have been as great as originally hoped; and natural precipitation variability might obscure significant results. Nevertheless, McDonald recommended "realistic optimism" and noted that the conference had made a strong case for continued cloud modification and cloud physics research.[57] In short, while everyone wanted to see the research continue, atmospheric scientists were having a difficult time providing reliable evidence that cloud-seeding techniques significantly modified clouds or the precipitation they produced.

The ACWC evaluation continued until summer 1957, when committee members sent Eisenhower a brief update of their accomplishments, stressing once again that weather modification research could ultimately "make a substantial contribution to the solution of the Nation's water resources problems." After providing a summary of their conclusions to date, they reminded the president that because Congress had turned down their request for funds, they would be unable to complete their evaluation and analysis and file a final report.[58] And why weren't funds appropriated? Because the Bureau of the Budget had determined that the ACWC had completed most of its work and all that was missing was that final report. Instead of spending money on the final report, Director Percival Brundage suggested that the administration compile information from earlier reports and turn the funding and responsibilities over to the National Science Foundation.[59] The ACWC submitted the information it had on hand.

The Final Report

Orville's letter and report—along with a "My dear Carl" letter from Lewis Douglas to Arizona senator Carl Hayden and a telegram protesting the lack of funding sent from Douglas to General Wilton B. Persons of the White House staff—must have a struck a nerve.[60] Eisenhower reassured Orville that the committee would receive sufficient funds to complete its final report, and on December 31, 1957, the committee submitted the first volume of the two-volume *Final Report of the Advisory Committee on Weather Control*. Volume 2 arrived less than a month later, minus the personal meeting with the president that Orville had sought on behalf of the committee.[61]

Volume 1 contained the conclusions and recommendations upon which all of the committee members could agree, while volume 2 provided technical reports with conclusions and recommendations that were "not necessarily those of the committee," as Orville wrote in his foreword. He also pointed to the "extreme shortage of competent scientists and engineers" in weather modification and the overall lack of basic knowledge of atmospheric processes. Because the committee held that weather modification was a topic of vital importance, members recommended that all of its records be turned over to the National Science Foundation, as the most logical governmental agency to develop long-range basic and applied research efforts. The committee endorsed passage of S. 86 (Eighty-fifth Congress), which would designate the NSF as the lead agency on weather control research and provide it with funding to carry it out.[62]

The committee had undertaken two major programs: the statistical evaluation program to determine if commercial cloud seeders had produced statistically significant (i.e., likely not due to random chance) increases in precipitation, and the physical evaluation program to determine the physical effects of silver iodide seeds on clouds, whether dispensed from ground generators or aircraft.

From the statistical evaluation, the committee found that seeding winter-type clouds in the West's mountainous areas produced a 10–15 percent increase in average precipitation that could not be attributed to natural variation.

However, it did not see this same statistical result in nonmountainous areas. The committee carefully pointed out that it was not implying that no effects were produced, just that the techniques it had used were unable to pick up the small variations that might have occurred. Nor did it find any evidence that cloud seeding intended to increase precipitation had led to decreases instead. It was also unable to determine the efficacy of attempts to suppress hail due to inadequate hail frequency data against which to compare the experimental data.[63]

From the physical evaluation, the committee concluded that silver iodide ground generators could produce ice crystals in favorable clouds, but the mere presence of ice crystals in the clouds was insufficient to produce precipitation. Additionally, earlier concerns that silver iodide seeds decayed rapidly in bright sunlight turned out to be unfounded, and seeds were found to be effective up to thirty miles away from the release site. It also found that sometimes clouds with temperatures that allowed ice crystals to form on silver iodide seeds contained very few, if any, naturally occurring ice crystals.[64]

Based on these findings—and the starting premise that scientific knowledge of physical and chemical processes of the atmosphere had to serve as the basis for weather control—the ACWC recommended a broad-based, long-range research program concerning these basic meteorological processes be undertaken by government, academia, and industry. This effort would require "vigorous" governmental funding and the encouragement of talented men (yes, only men) to enter the meteorology discipline. To carry out such a program, it recommended the NSF as the appropriate supervising agency. It also provided a laundry list of "deficiencies" in meteorological knowledge from the effects of solar disturbances on weather to the dynamics of precipitation processes, all equally important for numerical weather prediction, which had reached the operational stage in the preceding two years.[65]

With conclusions and recommendations about weather control in the hands of the president, Congress, executive departments, and atmospheric scientists throughout the country, the federal government was posed to make decisions about what it would do to make weather control a reality and to put it to work on behalf of the nation's economic well-being, its national defense posture, and its diplomatic efforts around the world. By early 1958, the Congress, the states, and meteorologists were all converging on a common solution to weather control inspired in large part by a Cold War–driven desire to one-up the Soviet Union in atmospheric science, which depended on a significant boost in research funding and scientific manpower. Operational weather control appeared to be within reach, or so some state officials thought. The question: if the state were "officials doing things," what official would grab weather control on behalf of the state? And once he had it, would this tool be powerful or would it wither away? After fifteen years of theoretical and applied research, the American state would finally control the atmosphere.

The cold war has become, in large measure, a technological race for military advantage.

DAVID Z. BECKLER (1955)[1]

The *Final Report of the Advisory Committee on Weather Control* serves as an appropriate dividing line between weather control's developmental period and the all-out federal attempt to control weather for domestic prosperity and security at home while controlling weather to curry favor with fence-sitting developing nations and fight proxy wars as the Cold War intensified. From 1950 through 1957, congressmen, state government officials, and meteorologists were forced to adapt to rapidly changing political, social, and scientific circumstances. After all, during this period *anything* looked technologically possible, and in many people's minds there was not much difference between science and technology/engineering. Whether they were eradicating mosquitoes or fire ants, or wringing every possible drop of water from the Colorado River before it crossed into Mexico, controlling nature was the order of the day. Legislators and bureaucrats at federal and state levels grappled with issues of state control and decision-making authority as they weighed the influence of "expert" scientific advice on matters of public policy, while meteorologists grappled with protecting their basic research agenda from governmental and commercial influences as they sought to maintain their newly won professional reputation.

Although federal legislators had originally aimed to fashion a weather control commission modeled on the Atomic Energy Commission, they ran into resistance from the military services and various camps of meteorologists. Military services were the primary patrons of weather control research, and they wanted no regulations that might interfere with their classified projects. The different camps of meteorologists—government (USWB), academic, and commercial—resisted for different reasons. Weather Bureau chief Francis Reichelderfer had no use for weather control when his underfunded bureau was struggling to get numerical weather prediction (computer model-generated weather forecasts) off the ground. Academic meteorologists, afraid of being lumped together with "quacks and frauds" selling rainmaking services in the US West, struggled to explain their lack of knowledge about precipitation processes. And commercial meteorologists, who had readily adopted the techniques pioneered by General Electric in the late 1940s, didn't want any regulations. Ultimately, the congressmen backed down, but during the hearings the meteorologists realized that the only expert testimony that mattered came from those associated with Irving Langmuir, the Nobel Prize–winning chemist who was positive he could control weather across the entire continent. In a game of "who ya gonna believe—some meteorologist or a Nobel Prize winner," the congressmen equated "prestigious" with "expert," much to the chagrin of the meteorologists.

After several failed attempts to set up a regulatory mechanism at the federal level, the congressmen settled for establishing the Advisory Committee on Weather Control. When its science-based final report recommended against federal regulations, they took a different approach, making the National Science Foundation the lead agency for weather control—a decision helped along by Cold War concerns over possible Soviet superiority in the field. This effort did not, of course, preclude the military services from launching or continuing their own efforts toward using weather as a weapon.

Worries over Soviet capabilities did not necessarily translate into consistent funding support for basic atmospheric research. As ACWC chairman Howard T. Orville noted after his committee disbanded, the "entire research effort can be seriously jeopardized by the whim of one or two public officials in permanent positions in government." Funds that had been "almost completely obliterated" in 1957, due to "economy-minded" officials, had been saved by the Sputnik launch. And yet the House Appropriations Committee disallowed the NSF's fiscal year 1960 request for $500,000 for the nascent National Center for Atmospheric Research and $2 million to carry out the research and evaluation of weather modification. "Such actions on the part of uninformed public officials," Orville fumed, "account for the weak, uncoordinated, and halting atmospheric research effort today."[2] While funding for basic atmospheric research did increase in the 1960s, the tension between basic and applied research dependent on federal funding remained.

Meanwhile, the states had taken regulation into their own hands. Although few states had enacted laws regulating weather control until the ACWC came into existence, by late 1957 thirteen states—including Washington—were regulating weather control to some degree; three states had passed related legislation; and an additional nine states had expressed interest in weather control. Some of the states had claimed sovereign rights to atmospheric moisture within state boundaries, and others had established their own research programs.[3]

Ten of the states regulating weather control, including Washington, required licensing. Did that dampen the weather wars? Not in Washington. Starting January 1, 1958, its Weather Modification Board began issuing licenses to commercial seeders, and the board also determined the competency of commercial operators, compiled detailed records of all weather control operations, and promoted basic R&D activities related to weather control.[4] Although the board thought it had done a reasonably good job of "getting the word out" about weather control operations, a good number of Washington residents were convinced that the continuous rains had to be due to cloud seeders, even if they lived far outside target areas.[5] Letters of complaint, which today provide hours of reading entertainment, were far less amusing to those at Washington's Department of Conservation who had to investigate each one and find a suitable answer. Some letters expressed concern that those living downwind of seeding areas might be deprived of rain ("Just because you are a state official, you certainly have NOT assumed the statis [*sic*] of GOD as YET"),[6] and some were concerned about all of the clouds blocking the sun ("Can this be considered healthful?").[7] Some farmers considered the artificial rain as tantamount to trespassing ("The law which creates the [Weather Modification] board, empowering it to issue licenses for rainmaking is unconstitutional. It denies our individual rights"),[8] and others scoffed that the required "public legal notice" of seeding was a joke when placed in newspapers like the *Cowlitz County Advocate* (circulation one thousand), published in the 1,400-resident burg of Castle Rock.[9] Even Governor Rosellini started getting letters, including one from a despairing resident concerned about the tourist trade. She wrote, "We are making an effort to attract tourists to our State, but how can we ever hope to successfully do so when the weather is so constantly miserable?"[10] Those who have camped in a soggy tent while visiting Mount Rainier can commiserate.

As the 1950s ended, Rosellini may have rued the day that cloud seeders entered his state when the original trickle of complaints became a flood. The weather war that pitted cherry farmers against wheat farmers continued in the lee of the Cascade Range, and the utility companies, which were seeding to enhance the snowpack for hydroelectric power, and the farmers and loggers who opposed them in southwestern Washington continued to do battle as well. In summing up the situation, University of Washington meteorologist

Phil Church thought that there had been fewer problems with snowpack enhancement, although the public tended to think *non-operational* silver iodide ground generators were able to modify the weather. People were, in his opinion, also lawsuit happy. "One man in this state," Church wrote to geophysicist Walter Orr Roberts, chairman of Colorado's Weather Control Commission, "now claims he has become impotent because of silver iodide fumes! What next?" What next indeed? Washington State University had determined that there was no significant difference between rainfall amounts in years with seeding and years without—a result viewed skeptically by local residents. Church was convinced that they needed to continue working on basic cloud physics projects, and he intended to focus his research in the lab.[11] At least his Seattle neighbors would not be upset about that.

How do the events discussed in the preceding three chapters fit into ideas about the state and its relation to science and technology during the Cold War? In military and legislative interest in weather control we see the developing symbiotic relationship between professional scientists and the state. The state needed the scientists' professional expertise and research skills to move weather control forward, and the scientists needed the state's patronage to fulfill their research agendas. This interweaving of science and state is what Brian Balogh defines as the "proministrative state." One of the curious claims that Balogh makes about the proministrative state is that it consistently "promised services that outstripped demand for them." Examples included state claims that nuclear energy would provide electricity so cheaply that it would not be worth metering, and state promises to "eradicate poverty."[12] Weather control fit right into that pattern. No one was demanding weather control in 1950. Irving Krick and his group were hawking their wares from the Pacific Northwest through the Intermountain West and into the Desert Southwest, but were people around the nation clamoring for designer weather? Not at all. But these three examples of the proministrative state have a common thread: control. Controlling the atom, controlling people's economic lives, controlling the weather. All for the better, of course. And all because in the post–World War II era, the nation had, as historian Michael Adas writes, "a faith in scientific and technological solutions, and a missionary certitude that the United States was destined to serve as a model for the rest of humanity."[13] Take that, Russian Bear!

A more science-focused federal program would not necessarily have made the meteorology community any more receptive to the possibilities of operational weather control, nor would it have stopped the weather wars. Its possibility did prompt heads of federal departments and agencies to jockey for position as they argued for making weather control a state tool under their supervision—recall our definition of the state, "officials in action"—while simultaneously expanding their power base. Whether it was turning the atmosphere into a water reservoir, preventing lightning from starting fires or hail from damaging crops, keeping fog from ruining airline passengers' days, or steering hurricanes "harmlessly out to sea," weather control had something for almost everyone. And the American state was going to make sure it discovered and used all of them to its advantage.

∗ III ∗

Weather Control as State Tool (1957–1980)

U.S. Seriously Concerned; Cold War May Spawn Weather Control Race

WASHINGTON POST AND TIMES HERALD, page 1 headline, December 23, 1957

YIKES! WE'RE BEHIND!

So screamed the front-page headline just three months after the Soviet Union's Sputnik launch had kicked the "space race" into high gear. Staff reporter Nate Haseltine drew readers in with the lede:

> The next hot fight on the cold-war front may well shape up into an all-out scientific race between this country and Russia to work out ways to control the world's weather. American scientists are seriously concerned that the Soviets may win the race, gaining a fair-weather monopoly for themselves and weather extremes for this continent.[1]

Whew! The source for this stunning announcement? The "still under wraps" *Final Report of the Advisory Committee on Weather Control.*

After describing some of the wilder schemes linked to weather control—melting ice caps with sprayed-on lampblack, leveling mountains to change entire climate patterns, towing ice south from the Arctic Ocean to change the temperature of the Atlantic and Pacific oceans—Haseltine looped back to what meteorologists already knew: meteorological knowledge was not sufficient to tackle them. As Weather Bureau chief Francis Reichelderfer put it, basic knowledge of atmospheric mechanics was "abysmal," but the Soviets were "speeding up and expanding basic research in meteorology" and training more meteorologists than the Americans were.[2] If government leaders hadn't seen science and technology as an affair of the state and the key to national prestige in the past, now was the time to get serious about providing sufficient support to enable their use as a state tool.[3]

Massachusetts Institute of Technology's Henry Houghton, hesitant to advocate for more than basic atmospheric research in the early 1950s, was worried about the unknown Russian threat: "I shudder to think of the consequences of a prior Russian discovery of a feasible method of weather control. International control of weather modification will be essential to the safety of the world as control of nuclear energy now is. Unless we remain ahead or abreast of Russia in meteorology research, the prospects for international agreements on weather control will be poor indeed."[4]

Testifying before the Senate Military Preparedness Committee in November 1957, famed hydrogen bomb scientist Edward Teller, who perhaps never met a large-scale weapon that he didn't like, had said, "Please imagine a world in which the Russians can control weather in a big scale where they can change the rainfall over Russia, and that—and here I am talking about a very definite situation—might very well influence the rainfall in our country in an adverse manner. They will say, 'We don't care. We are sorry if we hurt you. We are merely trying to do what we need to do in order to let our people live.'"

And rounding out the doom and gloom, Haseltine quoted the late John von Neumann, who had been convinced that numerical weather prediction would eventually lead to weather control: "Probably intervention in atmospheric and climate matters will come in a few decades, and will unfold on a scale difficult to image at present." Such an intervention would "merge each nation's affairs with those of every other more thoroughly than the threat of a nuclear or any other war may already have done."

"The next war clouds," Haseltine ominously summed up, "may be truly atmospheric."[5]

SOVIET AND CHINESE WEATHER CONTROL

Cold War national security had influenced weather control efforts since Project Cirrus, but the Advisory Committee on Weather Control (ACWC) report coupled with Sputnik brought that connection into sharper focus. The director of the CIA, Allen W. Dulles, was drawn in, telling his section heads that the White House was "becoming increasingly interested in the possibilities and potentialities for controlling weather." Recognizing this as a "vital subject," he planned to meet with the National Science Foundation's (NSF) Alan Waterman to get up to speed.[6] The CIA had been tracking international weather control efforts, including the Soviets', since the early 1950s.[7]

What were the Soviets doing? Writing in the journal *Priroda*, V. A. Shtal' and V. G. Morachevskiy argued for weather control research in support of the national economy, including rain creation, reducing or eliminating crop damage due to frost, dry winds, and hailstorms, and the dispersal of fog and low stratus. They discussed Western press accounts about using thermonuclear devices to control global atmospheric processes because anomalous weather phenomena seemed to follow weapons tests—a claim they disputed. Atomic bombs would not modify the weather, but local seeding would.[8]

Atmospheric scientist E. K. Federov claimed that the first Soviet cloud physics experiments got underway in the 1930s, the same time that Tor Bergeron and Walter Findeisen were doing their work (chapter 1).[9] Unable to make rain in quantity, researchers suspended their experiments during the war. The Soviet goal, he wrote, was to use weather control for good, while the United States used it for business purposes and to promote meteorological warfare. Credulous farmers had paid millions to entrepreneurs claiming to bring rain to their parched land. Political and military figures were openly discussing "meteorological warfare" and the alleged advantages that the Americans would have in a "meteorological attack" on the USSR—discussions that had "penetrated into the scientific literature." However, Federov continued, the "morbid speculation" on weather control had since subsided and widescale investigations were taking place in a number of countries.[10]

Since precipitation mechanisms remained undetermined, Federov reported, the Soviets were focusing on dissipating clouds while other countries sought to create precipitation, which could bring devastation. Indeed, "advancing deserts, ever-increasing soil erosion, dust storms, disastrous spring floods" were all present in the United States, which was exhausting its natural resources as it took a "predatory approach to natural wealth," while the Soviets worked to improve the natural condition by building reservoirs, drying marshes, and planting forests.[11] Not for them the West's "clumsy attempts to use the first scientific achievements on exerting an active influence on clouds for stirring up war hysteria," which were being pursued by "reactionary military and political figures in the USA and in certain other capitalistic countries." No, the Soviets would focus on scientific discovery and the improvement of natural resources, not their misuse and destruction.[12] Not exactly the tone one expects to see from a scientist, but Cold War references were not lacking in US scientists' pronouncements either.

Meteorologists attending the 1959 American Meteorological Society annual meeting heard J. Robert Stinson of the St. Louis University Technological Institute and Louis J. Battan of the University of Arizona Institute of Atmospheric Physics argue that the United States could be falling behind the Soviet Union in meteorology and weather control. Battan noted that Soviet research seemed similar to US efforts, but that "there was little available information about their work in the use of radar, cloud seeding, and weather modification." That lack of information must mean that it was secret, implying that what US meteorologists were doing was *not* secret, when much of it was. Stinson referred to Soviet meteorological research as "crude," but that some of it was "excellent by any standard" and often "bold," and with more investment in facilities and training programs, they would "catch up and eventually pass [the West] in research potential."[13] The solution? The United States needed to immediately move to recruit more meteorology students and support the science—yet another example of using the lack of knowledge of Soviet science to push for additional funding.

But Aleksandr M. Obukhov, the director of the Institute of Physics of the Atmosphere, Soviet Academy of Sciences, threw cold water on the overheated Americans when he expressed doubt about large-scale weather control and suggested that the best weather control device was an umbrella. Yes, the Soviets had undertaken small-scale experiments and achieved "moderate climate changes through modifying the landscape," but those were a long way from massive rainmaking projects, which would not be economical.[14]

And what of America's other Cold War nemesis—Red China? The People's Republic of China began weather control experiments in August 1958, following a severe drought in the Kirin region. Meteorologists noted that scientific and technical work under the Communist Party's leadership should be "primarily geared to the development of production and for service to socialist construction; and science should be motivated by 'missions,'" so meteorological research plans were accelerated to meet agricultural needs. The "City Committee" seeded nineteen times with several hundred kilograms of dry ice along a three-hundred-kilometer flight path, and the results were "very satisfactory."[15]

According to a review of the 1959 research year, Chinese meteorologists had made "magnificent accomplishments," and significant progress had been made in artificial precipitation. They had worked on aviation- and ground-based seeding, hail mitigation, and cloud and fog physics. Due to extant scientific and technological conditions, they had not achieved complete control over the weather, but had successfully modified clouds and fog. Of special note:

> Prior to the great leap forward, the Institute of Geophysics was actively interested in weather forecasting, which, of course, is still important, and an urgent need of national economic departments. But forecasting and analysis are not enough to satisfy the grand expectation of changing nature as demanded by the working masses during the great forward leap. We did not understand this point until the party made it clear to us, then we correctly decided to proceed rapidly with research in artificial precipitation.

Once in tune with the party, meteorologists faced sorting out the information published in the three thousand papers on cloud physics and weather control that had appeared in the preceding twenty years. Studying all the papers would take too much time, so they started operational weather control first and worried about cloud physics later. A "scientifically backward country" like China needed "high speed development." Hence, they started not with specialized research, but with the "masses who have been under the leadership of the local party and political authorities." Since the masses had been suggesting that they obtain more water and store it for agricultural and industrial use, it was important to engage the support of party leaders and seek the "active participation of local meteorological workers" from the start. The author continued, "The rapid progress during the past 12 months symbolizes the triumph that can be achieved when scientific work follows the line of the people."[16] Nothing like party encouragement to focus the mind.

According to an American assessment of the Chinese program, participants included personnel from the Institute of Geophysics and Meteorology of the Chinese Academy of Sciences, provincial meteorological bureaus, and the People's Liberation Army, which provided military aircraft and may have participated in the scientific efforts. While official government reports claimed a 70 to 80 percent increase in precipitation, the United States viewed those results as "misleading and probably exaggerated." However, there was little doubt that the Chinese would continue their efforts, and even moderate success could significantly aid their economy.[17]

If the Soviets and the Chinese were making major efforts in weather control, the United States was not about to be left behind, domestically or militarily. As the 1960s began, extensive programs to modify the atmosphere to produce the desired designer weather for whatever environmental problem was in play began to take hold. But whether domestic (chapter 6) or military and/or diplomatic (chapter 7), guaranteeing America's security with artificially induced precipitation was the ultimate goal behind state control of the atmosphere.

CHAPTER 6

Weather Control as State Tool on the Home Front

Man is still a very long way from shepherding the winds. But wisdom firmly suggests that mankind start now to train the shepherds.

SATURDAY REVIEW, 1966[1]

Drought conditions in the late 1940s and early 1950s had raised awareness of water shortages and distribution problems, and how weather control might alleviate them. Although Langmuir's claims of producing precipitation across the nation via periodic cloud seeding had failed to materialize, droughts in the 1960s renewed interest in artificially induced precipitation. Federal, state, and municipal agencies worked to perfect and deploy weather control technologies to alleviate water shortages, and to expand their areas of responsibility as well. By the 1960s, weather control had become a fixture in the US West, where states using federal grants tried to control their own programs. Most of the money came from the National Science Foundation (NSF), but the US Weather Bureau (USWB), US Forest Service (USFS), and the Department of Health, Education, and Welfare (HEW) were all interested in a number of weather control–related activities.[2] The agency most closely connected to developing water resources—Interior's Bureau of Reclamation (BuRec)—had not yet entered the weather control field, but within a few years it would become a major player as cabinet secretaries jostled for position at the congressional funding trough, and states seeking to enhance their water resources profiles fought for their share of support.

Federal agencies were facing a number of challenges related to atmospheric science research, including the lack of funding, manpower, and research coordination among governmental departments, which pursued unique agendas indicative of the fragmented nature of weather control policies. In addition, the political maneuverings of federal agencies illustrate the problems inherent in trying to control something that is uncontrollable. Occasionally, someone would raise concerns about potential conflicts between winners and losers in the pursuit of precipitable water, ecological issues such as the effects of modified weather on plants and animals, or legal issues surrounding lawsuits between governmental entities going after the same clouds. Yet these concerns were generally pushed to the periphery.

This chapter presents three case studies of state-controlled weather in the 1960s and 1970s: the US Forest Service's (USFS) small-scale Project Skyfire, to investigate the possibility of preventing lightning-caused forest fires in the northern Rockies; BuRec's nationwide Project Skywater, to exploit the water vapor flowing over the Continental United States as a new source of fresh water for thirsty land; and the joint Commerce/Navy hurricane-busting Project Stormfury, to keep hurricanes from smashing into the United States.

By the early 1980s, all of these programs were dead. The reasons behind their respective denouements, as well as those of the military-related programs discussed in chapter 7, will be addressed in the conclusion to Part III.

US FOREST SERVICE: SUPPRESSING LIGHTNING IN THE ROCKIES (1953-1979)

In the 1920s, while both were under the Department of Agriculture (USDA), the Forest Service and the Weather Bureau started a joint venture to reduce the loss of timberland to fire. Particularly in the West, forest fires are often sparked by lightning, and forest fires, regardless of cause, generally spiral out of control during dry atmospheric conditions, so foresters and weather forecasters teamed up to discuss how the USWB could be of assistance.[3] Within a few months, it was establishing fire weather stations in the West to keep foresters aware of potential fire dangers. Relative humidity was the most important element of such forecasts: the lower the humidity, the more likely a fire would be easier to start and harder to stop. Fire danger increased during long periods of low humidity, particularly if large amounts of fuel, that is, bone-dry trees and underbrush, were present.[4] By midyear 1926, the Agricultural Appropriations Bill carried an additional $18,000 (about $300,000 in 2015) to support fire weather investigations and forecasting.[5]

Working together, weather forecasters and foresters determined that fire weather forecasts needed at least four elements: precipitation types and amounts, humidity increases or decreases, wind speed and direction, and geographic location.[6] Lightning made little difference, particularly on the Eastern Seaboard, because most lightning storms were accompanied by sufficient precipitation to keep combustible material wet, or to put out resulting fires before they did much damage.[7] To determine the relationship between fire and weather, foresters at the Northeastern Forest Experiment Station explored the interconnections among weather conditions, moisture content, and fuel flammability. To assist them, the fire weather forecasters needed to increase their forecasts' lead time and reliability.[8] They also created "foolproof" instruments for the USFS that would mechanically record weather elements without much human intervention. The weather forecasters equipped all fire weather stations with thermoscreens (i.e., the "little white houses with the weather gear inside") containing thermometers, sling psychrometers for determining relative humidity, and a rain gage. Some stations got anemometers for measuring wind speed and direction, and a very few got thermo-hydrographs, which recorded temperature and humidity on a continuously moving piece of paper.[9] What they lacked was a standardized procedure for compiling the data, a problem rectified in 1930.[10]

After collaborating for two decades, USFS officials were attuned to new meteorological techniques. Curious about weather control's possibilities, especially creating rain to extinguish fires in remote national forests, they asked General Electric's (GE) Vincent Schaefer if it were plausible. Considering the bigger picture, Schaefer responded, "It may be that dry-icing will be equally or more beneficial to you by reducing or even stopping the formation of lightning in individual clouds." What a revolutionary idea! If foresters could control lightning, all fire suppression agencies would benefit. Once fires start, they may travel fifty to one hundred miles before stopping.[11]

Intrigued, USFS officials accepted Schaefer's no-cost offer to visit their Priest River Branch Station (Idaho) in 1948. Explaining that supercooling in clouds caused aircraft icing and lightning, Schaefer thought thunderstorm intensity depended upon the amount of supercooled and liquid water in the cloud along with the cloud's vertical extent and turbulence. If seeding could disrupt cloud growth, it could reduce the number of lightning strikes and forest fires.

Schaefer planned to photograph lightning storms as they developed over three separate "breeding spots" in the Kaniksu National Forest, capturing the life cycles of two separate storms. The next time two lightning storms occurred, he planned to seed one with dry ice, keep the second as an unseeded control, and photograph each with a time-lapsed camera. As he set up his equipment, a large thunderstorm occurred, but it was not locally generated, none of the individual cells developed fully, and two of them merged, making photography impossible. Although Schaefer stayed for another three weeks, no additional thunderstorms developed. As he ruefully noted, "As so often happens in research, and it seems to be particularly the case in weather studies, the start of a program is the signal for the cessation of the activities to be studied!" However, Schaefer thought that it would be possible in the future to modify lightning- and hail-producing clouds that led to "cloud bursts," destructive winds, and other destructive weather phenomena.[12]

Schaefer returned to Schenectady, but he did not forget about the lightning problem. In 1951, he proposed that the Munitalp Foundation, which supported basic meteorological research, fund a study of weather processes using mobile research teams that could travel around the country and observe atmospheric phenomena in different regions. By using time-lapsed motion picture equipment, researchers could watch as simple cloud forms developed into more complex structures that produced lightning, hail, and high winds, and invite "people who worked outside," for example, foresters, ranchers, farmers, and naturalists, to work alongside them.[13] Originally, Schaefer had suggested a research program focused on hail because of the tens of millions of dollars in crop damage it caused each year, but Munitalp opted to support lightning research on behalf of the USFS.[14] Yes, that is correct. A private foundation was funding a research project for a federal agency.

Skyfire, a basic research project on lightning-caused forest fires, got underway in 1953. Lightning-caused fires destroy valuable timber, damage watersheds and ecosystems, and render recreation areas unusable. In the early 1950s, lightning sparked about 7,500 forest fires each year, causing almost $25 million in damage ($250 million in 2015 dollars) and caused most forest fires in the US West, so reducing lightning would aid the nation. Skyfire had two long-range objectives: to gain a better understanding of the occurrence and characteristics of lightning storms and fires in the northern Rocky Mountains, and to investigate the possibility of preventing or reducing the number of these fires by applying weather control techniques.[15] Why the Rockies? Because 70 percent of its fires were caused by lightning. In one instance, lightning triggered more than four hundred fires in twenty-four hours, severely taxing fire suppression agencies.[16]

Researchers quickly discovered that a lot of preliminary work was needed before they could even think about tinkering with the atmosphere. The first requirement: determine the locations where lightning-caused fires were most likely to occur, the date and time of their occurrence, how quickly they were detected, how fast the fires expanded, how much ground they covered, and other related attributes. The researchers also needed excellent grounding in local weather and climate data, and cumulus cloud and lightning behavior. They also needed information about local clouds and lightning: lightning storm characteristics; typical cloud breeding areas; common lightning storm paths; and distribution of lightning storms over time and space. US Forest Service workers assigned to twenty-two specially selected lookout stations in Montana, Idaho, northeastern Washington, eastern Oregon, and northwestern Wyoming started this ongoing effort in 1953. Perched high above the national forests in their fire lookout towers, young "spotters" spent the summer, binoculars in hand, scanning the horizon for telltale puffs of smoke that signaled a fire and then radioing location information to fire-suppression teams. Since they were already in the towers, using fire spotters to record cloud types and sky coverage, instances of lightning, and resulting fires made a lot of sense.

The assembled data were invaluable to Project Skyfire coordinators. The spotters reported that while mountain ranges were typical locations for cumulus cloud development due to orographic lifting, as meteorologists would have expected, cloud development varied day to day depending on atmospheric factors. Therefore, researchers needed to conduct micro- and meso-scale analyses along with careful topographic studies. When Skyfire researchers plotted lightning data, they discovered that strikes tended to follow a distinct line through the forest, information that assisted future seeding efforts. They also noted that lightning storms in the Rockies could be classified as one of three basic types: local air mass; frontal; or high-level, fast-moving. But the storms exhibited much greater variation within types than expected, and their distinguishing features were quite fuzzy. Concentrating on storm behavior, team members developed the project's next stage: cloud seeding.[17]

In 1956, initial seeding experiments were not conducted in the northern Rockies, but in Arizona, where researchers could try out both airborne and ground-based seeding operations on growing cumulus clouds, analyze the seeds' effects, and test specially created equipment. The weather did not always cooperate. However, they observed postseeding virga (i.e., "rain streaks": drops that fall from clouds and evaporate before striking the ground) and changes in cloud bases' height and structure. On a few occasions, Skyfire's mobile radar was able to register activity within a cloud. Based on its tests, the research team determined that it could analyze cloud-seeding outcomes by using aircraft, radar, photography, nuclei measurement, and weather analysis.[18]

Following the Arizona experiments, the team moved to the northern Rockies, where it established an experimental site in the Lolo National Forest near the summit of the Bitterroot Range near the Idaho-Montana state line (fig. 6.1). After testing their seeding and analysis equipment, and conducting preliminary experiments, researchers determined the best placement of ground generators (fig. 6.2), the number of silver iodide seeds needed per second, the best

FIGURE 6.1. Skyfire experimental locations, 1957–1960. From Donald M. Fuquay, "Project Skyfire Progress Report, 1958–1960," Intermountain Forest and Range Experiment Station, Forest Service, USDA, Ogden, Utah, 1962.

flight paths for airborne silver iodide seeding (fig. 6.3), and how to overcome the problem of small planes being unable to reach the top of towering cumulus clouds during dry ice–seeding runs. For the latter, they settled on cutting through one edge of the cloud and releasing the dry ice there.[19]

Using the 1956 fire season to determine the best research design, seeding experiments began in earnest during the 1957 season and focused on ground-based seeding's effects on growing cumulus clouds. Looking for a large-scale seeding effect, the researchers chose a test area thirty miles long and fifteen miles wide in the Lolo National Forest. They continued testing new silver iodide generators and developing techniques that allowed for physical analyses of seeded clouds. To increase seeding opportunities, researchers seeded at every opportunity—as opposed to seeding as a result of a coin toss, for example—and compared the results of seeded clouds to adjacent unseeded clouds on the same day. Otherwise, they risked having insufficient data to analyze at fire season's end. Placing their ground generators in the center of Skyfire's cloud survey network, observers recorded meteorological conditions and photographed clouds as silver iodide seeds plumed up from the forest floor. Their mobile command center allowed them to keep track of events throughout the area (fig. 6.4).[20]

FIGURE 6.2. Project Skyfire acetone-burning silver iodide ground generator. From Donald M. Fuquay and H. J. Wells, "The Project Skyfire Cloud-Seeding Generator," Intermountain Forest and Range Experiment Station, Forest Service, USDA, Ogden, Utah, 1957.

FIGURE 6.3. Project Skyfire airborne silver iodide generator. From Donald M. Fuquay, "Project Skyfire Progress Report, 1958–1960," Intermountain Forest and Range Experiment Station, Forest Service, USDA, Ogden, Utah, 1962.

FIGURE 6.4. Mobile control unit for Project Skyfire; Don Fuquay, Vincent Schaefer, and Jack Barrows (left to right), 1957. US Department of Agriculture Photograph, author's collection.

That first summer yielded twenty-nine seeding days, thirteen during lightning storms. The researchers reached no definitive conclusions, but gathered sufficient data to continue the project. They needed to learn much more about storm electrification to understand the causes of cloud-to-ground lightning strikes and why some storms were more intense than others. Additionally, they needed to know a lot more about lightning storm characteristics than their first four years of data had revealed. Researchers also needed to uncover the characteristics of silver iodide and other seeding agents and how they acted as freezing nuclei. They needed to develop specialized instruments to measure and record the number of freezing nuclei in the atmosphere and clouds, and to measure and record lightning strokes. Add in the development of appropriate photographic equipment and improved silver iodide seeding generators, and researchers had more than enough to do between fire seasons. They would need to conduct experiments fire season after fire season until they had amassed enough data to determine with any degree of certainty if cloud modification would lead to lightning suppression and a concomitant reduction in lightning-caused forest fires in the Rockies.[21]

What did the researchers ascertain about experimental techniques and lightning fires? Due to local weather conditions dependent upon the larger synoptic situation and topography, releasing silver iodide seeds from ground generators was too unpredictable; they could not be used for controlled area experiments, so more expensive airborne seeding would be required. Furthermore, Skyfire's study of lightning storms revealed that in the study area, 50 percent of lightning storms yielded 0.1 inch of rain, and another 20 percent 0.1–0.2 inch of rain—basically, not much. "Dry lightning" was typical and contributed to the fire danger. Another 20 percent of the storms produced hail or graupel (i.e., "soft hail"), which did not aid fire suppression. Add it all up, and only 10 percent of the lightning storms produced more than 0.2 inch of rain.[22] It would take a lot more than that to extinguish a forest fire burning dry fuel.

Schaefer and his team also determined that many mountain storms were not as electrically active as thunderstorms in topographically flat regions. More than 50 percent of storms had ten or less ground strikes, while overall, the average thirty- to sixty-minute-long storm had fifty ground strikes. The high average resulted from the 5 percent of lightning storms that produced more than three hundred cloud-to-ground strokes, that is, these few storms were responsible for one-third of the total strikes. Since most had low precipitable water and a cloud base higher than twelve thousand feet, the falling rain evaporated just under the clouds, creating erratic winds that whipped up resulting fires. If they were going to prevent forest fires, researchers had to figure out a way to tamp down the number of lightning strikes.[23]

Experiments and data collection continued for several years. After the 1961 and 1962 fire seasons, researchers concluded that they needed to use more silver iodide seeds per experimental trial, and that there was a close association between the vertical height of the clouds and the number of lightning strikes.[24] By 1963, Project Skyfire personnel had assembled sufficient instrumentation, seeding equipment, and statistical techniques to start a long-range field test of the effects of cloud seeding on lightning. They attempted to discover how seeding affected physical and electrical structures of mountain thunderstorms as they continued to study the relationship between lightning discharge and forest-fire ignition.[25]

By the 1964 season, data indicated that massive silver iodide seeding of lightning storms led to a significant reduction in cloud-to-ground lightning strikes. To carry out this part of the program, researchers had to seed incipient storms with twelve airborne and twelve ground-based silver iodide generators and then observe, photograph, and record resulting lightning strikes. Not only were they looking for a reduction in strikes, they were trying to determine if cloud-to-ground strikes that triggered fires differed from those that did not. Their data suggested that fires were more likely to be caused by lightning discharges that had "long-continuing current" (LCC) portions. If that were true, could these types of discharges be altered by cloud seeding?[26]

The next experimental season provided more evidence that cloud seeding suppressed lightning. Over several summers, they had seen an average 32 percent reduction in cloud-to-ground lightning in seeded storms. The Skyfire team began planning a larger version of its experimental program to test the possibility of lightning-fire prevention through a ground-based cloud-seeding system operated in a forested mountain area. It developed a dual-burn generator that could pump out 100 trillion nuclei/second for 110 minutes, and continued working on a high-capacity generator that could produce even more seeds for longer durations.[27]

Further seeding tests meant to suppress the LCC phase of lightning strikes resulted not only in a statistically significant reduction in the numbers of strikes, but also in the duration of the LCC phase from 182 milliseconds to 115 milliseconds, which was thought to provide an important tool for managing lightning-caused fires. By 1970, Skyfire was also collaborating with the Bureau of Land Management to suppress fires by seeding clouds to increase rain over or near fires, a technique they used in Alaska.[28] The Skyfire seeding techniques were also used in 1970 when severe forest fires broke out in the Cascade Mountains (which extend from Northern California through Washington State)—once to reduce lightning and once to augment precipitation to aid fire suppression. The seeded areas had no additional fires, but researchers were unable to determine if seeding was the cause. However, the real purpose of seeding the fire zones was to convert the Skyfire research procedures into operational techniques that could be adopted more broadly by federal and state fire suppression agencies. Shortly thereafter, the Interagency Committee on Atmospheric Science approved the establishment of the National Lightning Suppression Project, effectively expanding Skyfire geographically and offering opportunities to assess its operational feasibility.[29]

Field tests continued through the early 1970s, but were suspended mid-decade, and the last half of the decade saw only the final data analyses. The USDA pulled all funding for weather modification out of its fiscal year 1979 appropriation request.[30] Project Skyfire and cloud modification for fire suppression were no more, but they left behind a treasure trove of data about lightning processes and their relationship to forest fires, a continuing problem that on average leads to more than ten thousand lightning-caused fires each year that burn some 4 million acres.[31] If climate change leads to drier forests and increasing amounts of dry lightning as occurred during the 2015 fire season, the number of fires and acreage burned could rise significantly in the future.[32]

INTERIOR'S BUREAU OF RECLAMATION: THE ATMOSPHERE AS WATER RESERVOIR

In the early 1960s, BuRec commissioner Floyd Dominy, chafing under geographical restrictions, was seeking justification for extending his water resources portfolio from the seventeen westernmost states to the entire nation. Continuing drought conditions coupled with advances in weather control techniques might provide excellent reasons. Tied by law to building dam and reclamation projects, if BuRec could exploit weather control as a solution to water shortages, it could expand its mission, power base, and funding. Unlike USWB personnel, who were interested in scientific discovery, BuRec staffers were interested in practical, engineering applications. They were unwilling to delay weather control until the atmosphere had given up its secrets on precipitation processes. The Bureau of Reclamation planned to use the atmosphere as a water reservoir, inducing rainfall over watersheds feeding its reservoirs in the West and extant reservoirs in the East, a plan that differentiated it from other government agencies. The tussles it sparked with the Department of Commerce exemplify the tensions that surrounded the state's domestic use of weather control.[33]

Reclamation Gets in the Game

The Advisory Committee on Weather Control's final report appeared in 1957, but attracted no attention from high-level BuRec personnel until spring 1961.[34] Once it did, BuRec quickly reached out to former GE researcher Vincent Schaefer, who saw the bureau as a "responsible, capable, and enthusiastic" entity that could take over the engineering facets of weather control. He hoped BuRec's knowledge of water resources combined with weather control would strengthen its long-term objective of enhancing the nation's water supplies.[35]

Within a year, Dominy and his senior advisers decided to pursue a research program focused on practical weather control methods that would increase the water supply, particularly in BuRec's reservoirs. The bureau awarded its first research contracts to the South Dakota School of Mines and the University of Wyoming for testing nucleating agents on cumulus clouds. Field researchers identified two similar clouds, seeded one, and held the second as an unseeded "control." By observing differences between the two clouds' behavior in these "markedly successful" experiments, they determined the effects of nucleating agents on the clouds, the types of clouds most amenable to artificial nucleation, and the lead time required to induce precipitation on a designated target. Based on data gathered by radar, time-lapse photography, and personal observations, BuRec reported that clouds precipitated when seeded, stopped precipitating when seeding stopped, and started again when seeding resumed. The control clouds did not precipitate. To reach a conclusion on seeding's efficacy, however, they needed to conduct many more years of field investigation.[36]

The Bureau of Reclamation's Washington, DC, headquarters staff's exuberant take on one season of experiments generated a hostile response from its Denver field office, which labeled the "markedly successful" appellation a "gross elaboration" of facts without scientific foundation. The Denver staff members had obtained cloud physics data from the experiments that did not support the overly favorable conclusion, leaving NSF, which was responsible for the nation's weather control research program, in the uncomfortable situation of being unable to answer questions about BuRec's experimental methodology.[37] The Washington, DC, headquarters staff, in turn, huffed that "some of the scientific brains out there still believe nothing can ever be accomplished in weather modification." It was not the Denver field office's prerogative to repudiate headquarters' public statements, which "detracted" from BuRec's efforts to maintain cordial relations with NSF and the USWB, where those "scientific brains" worked.[38] Skeptical NSF and USWB meteorologists, suspicious of BuRec's project, did not appreciate the hand-waving explanations of BuRec's experimental methodology and results. Future success, at least in the minds of those outside of BuRec, would depend on all interested agencies working together and sharing information. The Bureau of Reclamation's weather control project had just gotten started and was already digging itself into a hole with its meteorological colleagues.

Successful experiment or not, BuRec moved forward with its proposed weather control project by couching it as crucial for future growth in the US West, an area they viewed as a potential "food deficit area"—a deficit that could be ameliorated by water (see fig. 6.5). The population of the eleven westernmost states had quadrupled in the previous fifty years and was expected to triple in the next fifty. Irrigated agriculture had matched food production to population growth, but the abundance was narrowing. As western farmland

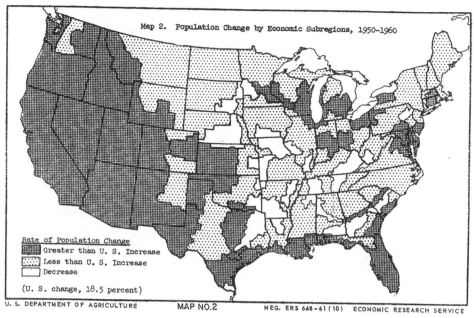

FIGURE 6.5. Population Change by Economic Subregions, 1950–1960. US Department of Agriculture, Economic Research Service. From "The West—A Potential Future Food Deficit Area," Bureau of Reclamation, 1963, Floyd Dominy Papers, American Heritage Center, Laramie, WY.

was transformed into housing developments that needed municipal water, the West would become unable to produce its own food by the early 1970s, and would produce less than half its needs by 2020 (figs. 6.6 and 6.7).

The Bureau of Reclamation calculated that the West would have to import almost 400 million tons of food annually to support its growing population, straining transportation systems and adding over 1 billion dollars per year to western families' food bills. During a national emergency, the situation could become dire. It was in the nation's best interests to ensure that the West could produce as much of its food as possible. The nation, according to BuRec, needed to step up its efforts to develop water resources to produce this food. The most readily available surface and subsurface water had already been developed or, in some cases, overdeveloped, and most local governments could not afford to develop remaining water. Federal and state governments needed to work together to meet the emerging need.[39] By tying its bid to enter the weather control business directly to food production, BuRec raised the specter of food production shortages within a decade in the US West. Implicit in

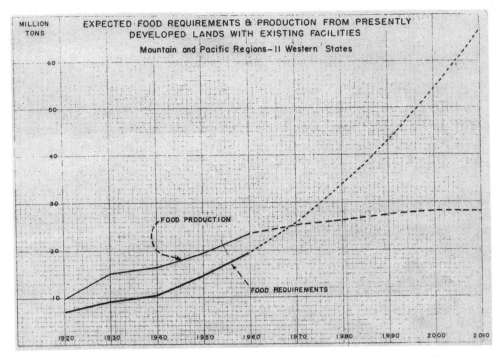

FIGURE 6.6. Expected Food Requirements and Production from Presently Developed Lands with Existing Bureau of Reclamation Facilities, Mountain and Pacific Regions—11 Western States. From "The West—A Potential Future Food Deficit Area," Bureau of Reclamation, 1963, Floyd Dominy Papers, American Heritage Center, Laramie, WY.

its argument was the classic Malthusian assertion that an agriculturally deficient West—a "third-world country" within an otherwise prosperous nation—would drain the nation's economy and subsequently weaken its defense posture. Although the public, and perhaps state government leaders, might have been convinced by this argument, more scientific-leaning members of the federal government were not.

Developing a Scientific Argument for Weather Control

One leader desiring more science and less posturing was Interior's science adviser, John C. Calhoun, Jr., who anticipated approaching NSF to fund BuRec's weather control program. The Bureau of Reclamation acknowledged that weather control faced significant difficulties: precipitation processes remained poorly understood, historical data were often inadequate, and commercial

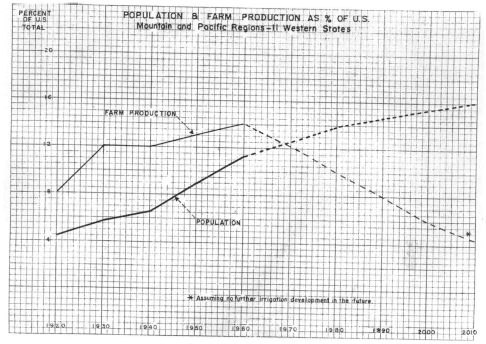

FIGURE 6.7. Population and Farm Production as Percentage of U.S., Mountain and Pacific Regions—11 Western States. From "The West—A Potential Future Food Deficit Area," Bureau of Reclamation, 1963, Floyd Dominy Papers, American Heritage Center, Laramie, WY.

rainmaking activities could interfere with scientific field programs. However, BuRec had decided on a different approach to weather control: seeding "abundant" atmospheric moisture flowing over its watersheds, focusing on resulting impounded runoff, and using an extensive network of instrumentation, e.g., radar and precipitation gage networks, and "photographs from U-2 aircraft" and satellites, to gather data. The Bureau of Reclamation planned to seek advice from NSF and the USWB before proceeding, but would focus on engineering and operational weather control, not basic research. It hoped to run the project over two eleven-year-long hydrologic cycles so that sufficient data would be available to "prove" it had altered the weather or not.[40] Its proposed project would tackle the applied side of weather control: producing precipitation over a specific watershed during optimal seeding conditions. The National Science Foundation and its other grant recipients would determine how and why weather control worked.

Calhoun was dismayed by the "plan"; it provided nothing he could use to gain support from NSF. Increasing water when moisture was abundant? Too vague. How would BuRec know if the experiment had been successful? What kinds of projects would it undertake? What types of clouds would it seed? Was it going to use a variety of techniques or just one? How would it address its lack of precipitation process knowledge? Calhoun was flummoxed. Why did BuRec think it was acceptable to pursue its own applied path instead of cooperating with organizations that had been, and would continue to be, conducting basic research with NSF funding? Seeking practical results alone would never meet NSF's requirements.[41]

Calhoun was particularly upset because he knew that the University of Arizona's weather modification studies had not produced detectable rainfall, and the University of Chicago's two years of experiments had produced insufficient data to develop a firm conclusion. Ongoing research aimed to understand small-scale (down to molecular) processes, and although accretion was important to precipitation processes in the lower and middle latitudes, no one understood how those processes happened so quickly. Basic and applied weather control research had to be coordinated. Only the most carefully planned field experiments would yield statistically significant results.[42] Bureau of Reclamation staffers were going to have to develop a scientifically solid program in order to get Calhoun's support, much less NSF's.

Regrouping, BuRec staff members conceded there was no statistical evidence of increased precipitation due to cloud seeding. However, they continued to argue that there was a difference between statistical significance based on precipitation recorded in rain gages, which might not be in the "right place," and evidence from stream runoff, which was BuRec's bottom line. They cared not one whit about the statistical significance of randomized cloud seeding because a precipitating cloud did not necessarily lead to measurable runoff. Therefore, BuRec proposed to seed every moisture-bearing cloud formation over a watershed, measure resulting streamflow, and compare it to a similar nonseeded watershed. Staffers were convinced that if they did not seed all moisture-bearing clouds, they would not increase water supplies.[43]

Staff members at BuRec quibbled over randomized cloud seeding, but not over the need for instrumentation. Bidding for Calhoun's backing, Dominy agreed to broker an agreement between Interior and NSF that would allow BuRec's Denver field office to prepare a detailed weather control research program, tossing in a few meteorologists to placate the scientific community.[44] Prepared to move "full steam ahead," BuRec opted to target one of its most important watersheds: the Colorado River Basin. However, it needed additional funds, and members of the Subcommittee on Irrigation and Reclamation of the Senate Committee on Interior and Insular Affairs, some of whom had been pushing weather control legislation since the early fifties, were ready to help.

Reclamation Finds Friends in the Senate

The Bureau of Reclamation got its desired fiscal infusion courtesy of Senators Henry M. "Scoop" Jackson (D-Washington) and Clinton P. Anderson, who called for hearings to determine the feasibility of its proposed program. As the hearings opened in May 1964, Anderson mentioned that a water consulting firm had reported that six times more water flowed above the United States as water vapor—some of which traveled over the entire continent without falling to the ground—than was carried by all of the nation's rivers. Wasted water! If this moisture fell on arid and semi-arid land that could subsequently be developed, it would have huge economic value. Therefore, Anderson proposed state support for two parallel programs: one of basic research into cloud physics mechanisms and another of extensive field research devoted to practical applications, the first "faltering step toward the ultimate solution of our water problems."[45] This sounds suspiciously like the hearings in the 1950s. Ten years after Anderson's earlier foray into weather control, his rhetoric remained unchanged.

Bureau of Reclamation and Interior staff members testified that weather control offered big opportunities for increasing water flow that would improve the environment and quality of life, presumably for people and not the drylands' extant inhabitants. With additional research leading to practical applications, weather control could significantly affect human activities—again, no mention of other life forms—and BuRec was just the agency to coordinate this effort. Now for the "ask": it needed more people, cooperation with other government entities, and, of course, money.[46] In the witness chair, Calhoun endeavored to ensure that the senators understood the extent of the required experiments. Perceiving his hesitation, Anderson chided Calhoun and challenged him to discuss possible downsides. Calhoun cautioned that weather control was not necessarily harmless. Overseeding might decrease rainfall, and seeding might contaminate the atmosphere and water. Atmospheric processes could become so skewed that scientists would lose the opportunity to gain vital knowledge about natural processes.[47]

Bureau of Reclamation commissioner Floyd Dominy presented a specific justification for weather control: its reservoirs had been built based on early twentieth-century historical streamflow, which was no longer adequate to meet demand. Increasing precipitation would increase streamflow and the quantity of stored water. The Colorado River Basin was already in a diminished water situation, so it provided an ideal opportunity to test the efficacy of weather control for water resources purposes. Although scientists could not agree on whether past programs had increased or decreased precipitation, no one could induce precipitation from dry air. Effective programs would enhance falling precipitation, which would land in watersheds feeding BuRec's reservoirs. Dominy testified that he had heard that "water engineering involves a combination of mechanical and theological skills." It was time to reduce the need for the latter. If BuRec could pursue its plan for a ten- to twenty-year program for $1 million to $2 million per annum, it should be able to increase the percentage of mechanical skills involved.[48]

The generally supportive senators asked some tough questions. Did BuRec have the authority to work on weather control? How would it handle the extra water? Why did extra precipitation matter when BuRec just let it evaporate from the surfaces of its huge reservoirs? Good question . . . Would the proposed methodology work in other places, for example, in the Missouri River watershed, which was expected to run out of water in 1980?[49] If successful and replicable, many more senators would be interested in "bringing home the water." How would senators know whether cloud seeding had worked or not?

Since BuRec did not plan on randomizing clouds, it considered randomizing watersheds: seed in one watershed and not in a similar watershed, and then compare differences in runoff and streamflow. Determining differences in snowpacks that would melt later might be difficult (depth alone is not a valid indicator because water content varies among snowfalls), but they would do whatever necessary to make detailed measurements and observations and thus develop a science-based conclusion. Bureau of Reclamation officials thought they might be able to get assistance from "defense agencies," but provided no details.[50]

Subcommittee members took the bait, and in late August 1964, Congress appropriated an extra $1.1 million to BuRec for the establishment of the Office of Atmospheric Water Resources in Denver.[51] In a joint announcement, Interior and BuRec maintained that the expanded weather modification research program would investigate the "potentialities of atmospheric water resources" with the help of academic institutions and consulting meteorologists. According to Dominy, the project would provide data to support the "growing conviction that man can influence the weather under certain conditions when sound engineering and scientific research techniques" are applied. It would initially focus on augmenting precipitation over watersheds feeding reclamation projects in chronically water-deficient locations.[52]

By late 1964, BuRec had a foothold in the federal weather control effort. As NSF, and to a lesser extent the USWB, pursued related cloud physics research, BuRec took on the engineering aspects of weather control to boost its reservoirs' water supplies. Would it maintain that focus? Or would BuRec look for ways to expand its influence and control over those who would control the atmosphere?

Consolidating Control

Bureau of Reclamation officials had "heard through the grapevine" that NSF intended to divest itself of weather control responsibilities in early 1965, and they promptly began angling to solidify the bureau's role as a major player in state weather control. Over the next two years, Floyd Dominy and his staff wheedled, cajoled, and politicked the Interagency Committee on Atmospheric Sciences (ICAS), NSF, the Department of the Interior, and the Bureau of the Budget in their attempt to take over the nation's scientific leadership of weather control. One small problem: ICAS's chairman was Undersecretary of Commerce J. Herbert Holloman, the Weather Bureau belonged to the Department of Commerce, and the chances that Holloman would allow BuRec to take regulatory control of state weather control were slim to none.[53]

Nevertheless, BuRec pushed for a broad-scale engineering approach as it developed a comprehensive strategy that would supplant existing "haphazard" operations, even as the Forest Service and Weather Bureau—who didn't think BuRec had any chance of taking over weather control—continued to offer it suggestions.[54] The Bureau of the Budget (BoB) was trying to ascertain the effectiveness of the federal government's various weather control activities, while the National Science Board was doing likewise.[55] The query that BoB sent out to various departments and agencies in its quest for information about weather control indicates that there was a major problem with domestic weather control: multiple purposes and paths with no one in charge, which encouraged political maneuvering for budgetary and scientific control. The bureaucratic fragmentation that was endemic in the American state was mirrored in weather control.

While BoB was sending out query letters, BuRec was sending Interior a military-allusion-filled strategy for its Atmospheric Water Resources Program. "We have," BuRec staffers wrote, "our own troops, allies (other agencies), Hessians (contractors), and a persistent enemy (water shortage)." They were assessing the situation, defining their objectives, and preparing orders for field units, all justified by congressional support and "somewhat controversial" indications that seeding clouds in orographic systems was effective. Bureau of Reclamation officials claimed they would establish a physical understanding of precipitation mechanisms while simultaneously developing engineering techniques for cloud nucleation to increase water supplies. Once these goals had been accomplished they would commence routine cloud seeding.[56] The Bureau of Reclamation expressed more confidence in its ability to create knowledge of precipitation mechanisms than did atmospheric scientists, who had been researching them for decades. Once again, those furthest away from weather control's scientific core were those likely to see inherent problems as simple to solve.[57]

Although NSF's Earl Droessler, who headed its Atmospheric Sciences Program, had been skeptical of BuRec's efforts, he was favorably impressed with its latest weather control strategy. Bureau of Reclamation personnel strongly argued that its longtime use of weather and climate records and of precipitation and runoff data demonstrated its atmospheric science capabilities, which was rather like claiming that having a rain gage in the backyard and a barometer in the house made one a meteorologist. (Or as a historian once told me, "I know as much as any meteorologist. . . . I watch the Weather Channel every day!") Nevertheless, Droessler thought multiple agencies should be attacking weather and climate control problems, "among the most exciting and promising facing scientists and engineers today," within their mission areas.[58]

With Droessler's support in their back pockets, BuRec officials were hoping to claim the upper hand in collecting weather modification data at the upcoming ICAS meeting.[59] No matter what ICAS decided, however, BuRec anticipated that its water resources program was going to get bigger—much, much bigger. In a "blue envelope letter" to Dominy, Chief Engineer Barney Bellport anticipated that its weather control budget would rise to $35 million or more *per year*, the related staff would number eighty to ninety people, with sixty in atmospheric water resources, and contract personnel would number one thousand, increasing eventually to 2,800.[60] With continued congressional interest in weather control, a recent weather modification conference, and a forthcoming conference on desalination, BuRec personnel thought the time was right to sponsor an international conference on atmospheric water resources within the next couple of years to give their agency "unquestioned" global leadership in the field.[61]

From late 1965 into early 1966, BuRec fielded requests from senators interested in expanding their program. The NSF and the National Academy of Sciences issued favorable reports on weather control, and both suggested that one agency needed to be in charge while recognizing that BuRec still had a mission in the weather control field.[62] It was time to determine the next steps.

The tentative plan called for BuRec to ask for a budget of $1–$2 *billion* ($7–$14 billion in 2015 dollars) for atmospheric water resources research over a decade, with annual budgets ranging from $20 million to $50 million in 1968 to $120 million to $140 million in 1977. Staffers envisioned six to eight regional centers (fig. 6.8) across the Continental United States that would carry out major research responsibilities within their assigned geographic areas, while other projects would cut across regions and require coordination. Considering the large amount of money involved, they would need to secure additional appropriations and figure out how to keep from sharing it with the Environmental Science Services Administration (ESSA)—the new umbrella organization including the National Weather Service and the Coast and Geodetic Survey.[63]

Another plan was submitted to Interior in late 1966. It differed from early versions because in his endorsement, Interior's assistant secretary of water and power development, Kenneth Holum, called for considering weather control's sociocultural impacts and its effects on nature. Interior needed to

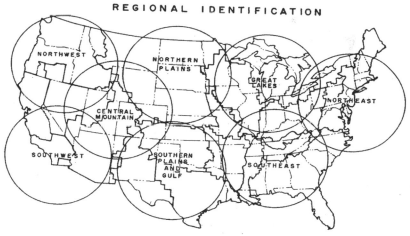

FIGURE 6.8. The Bureau of Reclamation's Atmospheric Water Resources Plan included dividing the country into regional centers, shown here. From "Plan to Develop Technology for Increasing Water Yield from Atmospheric Sources: An Atmospheric Water Resources Program," US Department of the Interior, Bureau of Reclamation, 1966, Floyd Dominy Papers, American Heritage Center, Laramie, WY.

develop a clear picture of environmental consequences, and to continue legal studies of weather modification along with scientific and engineering R&D. Since the plan called for developing an operational system for precipitation enhancement in some regions by 1972 and a general capability to enhance or redistribute precipitation by 1985, sociocultural and legal issues could not be put off much longer. Holum noted that President Johnson had already called for stepped-up efforts to make weather modification a reality, and members of Congress, especially Senators Anderson and McGovern, were similarly interested. Secretary Udall signed off in November 1966.[64]

The final plan had two parts. The first presented a snapshot of the nation's current and potential water deficiencies (fig. 6.9). Complacency was not an option. The Colorado River Basin was already deficient, and projected water needs for 1980 would not be met by the Missouri River, Western Gulf, Great Basin, and South Pacific watersheds if population growth and use trends were not altered or new water sources went untapped. The migration of people into water-poor areas was not just a western problem because most of the country was experiencing some level of water supply problem. If these areas had access to more water, agricultural, industrial, and recreational opportunities would increase. Water quality was important, too. Unless the United States better managed existing supplies, improved distribution, and increased the total supply, its future development would be hobbled. Interior could not just manage surface and subsurface water; it had to develop it. The atmosphere had yet to be adequately developed as a water source.[65]

The second part of the plan addressed how weather modification could change the nation's water resources picture. The Bureau of Reclamation needed to advance weather modification to increase precipitation to solve water shortages triggered by drought or distribution problems. Existing methodologies relied on basic and applied research in cloud dynamics, microphysics, and cloud-environment interactions that had been underway for two decades. But heading into the mid-1960s, a shift in attitude among scientists made effective cloud seeding appear to be more plausible as a water source. This shift had not come due to clearer theoretical insight, but due to the availability of more experimental data.[66]

The Bureau of Reclamation estimated that on average, the Continental United States received 4.2 trillion gallons of water per day in precipitation. Therefore, the water vapor in the atmosphere was a tremendous water reservoir that could be tapped by weather control techniques without materially affecting total atmospheric moisture. Long-term, BuRec wanted to focus on areas that needed water, not areas where it was simple to enhance precipitation,

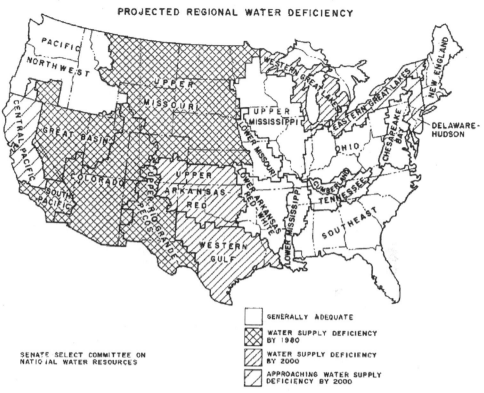

FIGURE 6.9. Potential water deficiencies based on data from the Senate Select Committee on National Water Resources. Note that BuRec anticipated that the Southeast would have generally adequate water supplies, an estimate that woefully underestimated a rapid increase in Atlanta's population and a concomitant drop in stream flow in the Apalachicola-Chattahoochee-Flint water basin by the early twenty-first century. From "The West—A Potential Future Food Deficit Area," Bureau of Reclamation, 1963, Floyd Dominy Papers, American Heritage Center, Laramie, WY.

such as mountainous areas with moisture-laden flow or lake-effect precipitation zones.[67] The program's name: Skywater.

Skywater

Skywater was openly billed as an applied weather control project to augment the nation's water supply in the shortest time. (See fig. 6.10.) The amount of water produced would provide the scientific answers that BuRec sought as it developed effective techniques and operating systems with a goal of increasing

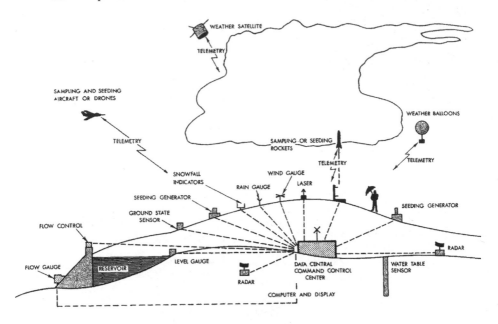

REPRESENTATION OF THE POTENTIAL ELEMENTS OF REGIONAL OR SUB REGIONAL ATMOSPHERIC WATER RESOURCES SYSTEM

FIGURE 6.10. Project Skywater's high-tech vision of the future of weather control, including automatic command and control systems, and drones for sampling and seeding clouds. Note the man carrying the umbrella—ready for falling precipitation. From "Plan to Develop Technology for Increasing Water Yield from Atmospheric Sources: An Atmospheric Water Resources Program," US Department of the Interior, Bureau of Reclamation, 1966, Floyd Dominy Papers, American Heritage Center, Laramie, WY.

precipitation by an average of 10–20 percent annually by 1985. This new, "inexpensive" water would be important for supporting social and economic growth in areas affected by water shortages and/or high water management costs. Increased streamflow and heightened soil moisture that resulted from seeding would help to meet water-dependent agricultural, industrial, and municipal needs. Additional streamflow would enhance water quantity, aid pollution control, and help to meet increased demand for water-based recreation. With an expected cost of $1–$4 per acre-foot and benefits ranging from $5–$50 and more per acre-foot, the benefit-cost ratio was anticipated to exceed 10:1.[68]

Skywater also included plans to study sociocultural impacts, economic effects, and ecological effects of weather control. Researchers would seek to "identify and resolve" problems that might result from the introduction of weather control; failing to do so could prove to be more of a limiting factor than technical or scientific problems. Social and cultural studies would be important for enhancing understanding and acceptance of weather control among the populace, as would attention to the effect of additional clean water supplies on the quality of life. Ecological studies would address short- and long-term effects of additional precipitation on fish and wildlife, microbioclimate, and vegetation. Legal studies would investigate problems related to liability, legislation, regulations, and authority. But despite the plans for sociocultural studies, they did not appear in BuRec's description of its studies in 1967. Legal and ecological studies, however, were undertaken at the University of Arizona and the University of Michigan (although there was no budget line item for Michigan's research).[69]

As Skywater gained momentum, Commissioner Dominy approached NASA administrator James E. Webb for assistance, arguing that space technology could be of major importance in solving the atmospheric water resources problem. Although the funding comparison was not even close, BuRec needed to make technological improvements that reflected those of the space program. Dominy's staff needed sensors, telemetry, computing capabilities, aircraft instruments, and delivery systems to identify suitable seeding situations and techniques for seeding, and to evaluate seeding operations in water-deficient locations in which social, legal, economic, and climatic conditions made weather control an acceptable means of providing additional water. Dominy urged Webb to join him in a collaborative effort to bring operational weather control to fruition.[70]

Noting the possibility of "modestly" affecting precipitation within a decade, two University of Michigan researchers conducting ecological studies for BuRec argued that broad-scale weather control would not be feasible in the near term or ever. They argued that a generation earlier, weather control technologies would have been applied as soon as possible had they promised substantial profits, but public attitudes toward the environment were in flux. Profit alone would not be reason enough to pursue weather control if it might significantly (and adversely) affect the environment.[71]

Weather control would be desirable only if weather changes resulted in a net economic gain coupled with adequate compensation for those who suffered economic losses, as well as a net improvement in psychological satisfaction for people and in environmental conditions for plants and animals. The authors contradicted weather control supporters' claims that since Western civilization had always been based on exploiting resources and modifying the environment, changes that provided an economic benefit were acceptable because technological fixes were possible when ecological troubles surfaced. But the atmosphere varied from year to year, making it more difficult to determine whether changes in plant and animal communities were due to artificially induced weather or to unexplained changes that would have taken place anyway.[72] While the federal government may have viewed the "environment" in terms of the physical environmental sciences and how they served the nation's defense, those working in ecology and natural resources looked at it with a plant- and animal-eye view. They would not always see more water as beneficial. For BuRec, and industries, agriculture, and municipalities in water-poor areas of the country, more water was always an advantage.

The Bureau of Reclamation considered Skywater and its goal of augmenting the usable water supply from "rivers in the sky" as being yet another technological fix in a long line of similar water supply–fixing technologies: harnessing rivers, drilling wells, reservoir and irrigation canal systems, and desalination.[73] Although some people were concerned that Skywater's seeding experiments could lead to flooding if too much snow accumulated in watersheds, program managers argued that they were not trying to produce record snowpacks. They would seed until the snow reached a predetermined depth, and only in areas that possessed adequate reservoir storage. Contrary to arguments that they might upset the "delicate balance of nature," BuRec officials argued that this was not a delicate balance at all. Nature was not benign: tornadoes, hurricanes, and severe thunderstorms caused millions of dollars of damage annually. Skywater, which would gather moisture that would otherwise float overhead, would be of the utmost benefit to society.[74]

The Bureau of Reclamation saw no downsides from Skywater. The tremendous amounts of moisture carried in the atmosphere meant that seeding in one area would not deprive those downstream of water. Seeding with silver iodide was not an ecological problem because it was already less than 1/100th of the level permitted by the US Public Health Service. Furthermore, seeding was already underway in twenty-five states and in many countries. The Soviet Union's weather modification program was two to three times larger than the United States'. BuRec then played the famine card, noting that increasing populations around the world demanded more food. America's strength would be determined by its ability to manage the rivers in the sky, and feed not only its own people but also others in need around the world.[75]

The Bureau of Reclamation's arguments for Skywater boiled down to one: national security. The United States' Cold War nemesis, the Soviet Union, was outspending it on weather control. Controlling the weather meant the difference between secure and insecure food supplies for the United States as well as for its friends, foes, and fence sitters in a world divided between the East and West blocs. If the Soviets pulled ahead in the weather control race, they would be able to solve their agricultural problems and become a more formidable adversary. They would also be able to extend their weather control techniques to non-aligned countries and pull them into their ideological camp. Since the Departments of State and Defense were on hand to make the claims that being behind in weather control threatened American security, BuRec did not have to make these arguments directly. However, it could allude to falling behind America's Eastern adversaries as a selling point for weather control. Thus BuRec's arguments were aided by a Cold War mentality, the continued attractiveness of technological fixes, and an ongoing drought that deepened in the mid-1960s. In recent years, these arguments might have seemed overwrought as California alone continued to produce almost half of the nation's homegrown nuts, fruits, and vegetables instead of waiting for life-saving food shipments from the Eastern Seaboard. But in the 1960s, with widespread fear of environmental degradation and discussions in California of dividing the state into the "well-watered" North and the "water-deficient" South, the sky-is-falling-in prediction of a highly populated area running short of food and water made perfect sense to those holding the nation's purse strings. Although no one is anticipating that California will become a net importer of food in the near future, the deepening drought in summer 2015 led to state government cutting off water supplies to farmers for the first time ever. That decision will adversely affect agricultural output, but even without state restrictions, farmers would not have found enough water to keep their water-intensive crops growing. As of July 2015, the emphasis has been on statewide conservation, and weather control efforts seem minor by comparison. The blue skies are lovely and the balmy temperatures a treat, but cloudless skies do not provide an opportunity for seeders to wring water from the atmosphere.

COMMERCE'S USWB: SNUFFING OUT HURRICANES WITH PROJECT STORMFURY

The idea that people could one day prevent hurricanes from forming, snuff out little hurricanes before they grew to enormous proportions, or, failing that, steer them away from land and send them "harmlessly out to sea" had been bandied about for a while before the discussion of computer-generated weather forecasts and their promise of controlling the weather first hit the *New York Times* in January 1946:

Some scientists even wonder whether the new discovery of atomic energy might provide a means of diverting, by its explosive force, a hurricane before it could strike a populated place.[76]

Apparently having radioactive fallout scattered about the world's oceans from one-hundred-plus miles per hour winds was not a concern, but the use of atomic weapons to bomb hurricanes out of existence did not seem so far-fetched in the late 1940s and into the 1950s when the United States was conducting aboveground atomic tests in the Pacific Ocean. Even today, a NOAA website addresses this possible use of nuclear weapons in its "frequently asked question" section.[77]

Meteorologists of the USWB were not among those suggesting ways to snuff out hurricanes, since very little was known about hurricane structure, dynamics, and behavior in the immediate postwar years. That had not stopped Irving Langmuir from developing a plan to seed a hurricane, preferably a "young" one, in concert with his military funders as part of Project Cirrus. He wanted to gain experience in seeding tropical cyclones that might be used to better steer or otherwise modify them in the future. During the 1947 hurricane season, General Electric assembled a team of consultants, military aircraft were maintained on "stand-by," and the wait for a suitable tropical system began. With the hurricane season winding down, word came on October 10, 1947, of a hurricane that was forming in the Caribbean Sea. Within twenty-four hours, the Project Cirrus B-17s had arrived in Mobile, Alabama, but the fast-moving storm was already crossing Florida, so the next day the planes headed for Mac-Dill Field in Tampa, Florida, to join the Air Force's 53rd Weather Reconnaissance Group. The combined units intended to take off on the seeding mission the morning of October 13 and head for the hurricane, which was expected to be between 480 and 640 kilometers east of Florida.

Instead of attempting to penetrate and seed the eyewall, the research team seeded the uppermost cloud shelf far enough from the center that reconnaissance aircraft could photograph the effects of seeding from 1.5 kilometers above the seeding aircraft. The first seeding run took thirty minutes as thirty-six kilograms of dry ice were distributed along a 180-kilometer track. Then the seeding aircraft dumped two twenty-two-kilogram loads of dry ice into a large cumulus cloud. Seeding accomplished, the aircraft retraced their paths, taking visual and photographic observations. According to Lieutenant Commander Daniel F. Rex, a Navy meteorologist who later became deeply involved in numerical weather prediction, the seeded cloud deck showed "pronounced modification," with the previous overcast area turning into "widely scattered snow clouds" over 770 square kilometers. The observers did not notice any convective activity (development of large cumulus clouds) following the seeding.

Assessing the outcome, Vincent Schaefer noted that this was an "old storm," instead of the desired "young storm." He concluded that a young storm would be less "complex," and researchers would have an easier time detecting seeding's effects. However, GE's home base of Schenectady was not within quick striking distance of developing tropical storms, and Schaefer thought he and his colleagues should desist from hurricane work until they had accomplished more basic work. Langmuir thought they needed to learn a lot more about hurricanes and continued, "It seems to me that next year's program should be to study hurricanes away from land, maybe out considerably beyond Bermuda, out in the middle of the Atlantic. . . . I think the chances are excellent that, with increased knowledge, we should be able to abolish the evil effects of these hurricanes."[78]

End of story? Not quite. After the seeding experiment was over, the hurricane's path changed from its northeasterly course to westerly (fig. 6.11)—right into Savannah, Georgia, causing some $2 million damage (1947 USD) and one death.[79] A decade later, meteorologists reconstructing the event provided evidence that seeding had not in any way influenced the hurricane's movement.

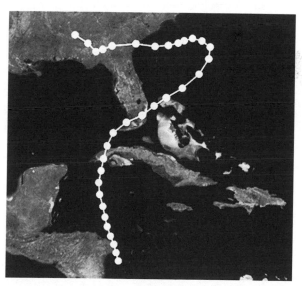

FIGURE 6.11. The track of 1947 Atlantic hurricane no. 8, which was seeded using Langmuir's techniques. Wikimedia Commons; URL: https://commons.wikimedia.org/wiki/File:1947_Atlantic_hurricane_8_track.png.

But at the time Langmuir took "credit" for the course change, much to General Electric's chagrin.[80] As his GE coworker, Bernard Vonnegut, would point out over twenty years later, Langmuir's "scientific vendetta with the hurricane did not stop"; he continued to call for a "comprehensive hurricane seeding program" whenever he gave a talk on weather modification.[81]

Atmospheric scientists' expressed desires for more knowledge of hurricane dynamics and structure would not have prompted the establishment of a national hurricane research program, but when the 1954 and 1955 hurricane seasons produced $2.15 billion ($19 billion in 2015) in damage and four hundred deaths in the United States from Hurricanes Carol, Edna, Hazel, Connie, Diane, and Ione, that seemed to call for state action. A congressional appropriation in spring 1955 provided the USWB with significantly more funding for hurricane research, and once the 1955 hurricane season ended, Congress told the USWB to establish the National Hurricane Research Project (NHRP), which ultimately became the Hurricane Research Division of the Atlantic Oceanographic and Meteorological Laboratory in 1983. Within a year of its establishment, the NHRP was using Air Force aircraft to collect data, and within six years it had its own aircraft and had produced almost sixty research reports. Its mission was to study hurricane formation, structure, and dynamics, "seek means for hurricane modification," and improve hurricane forecasts.[82] Using aircraft, radar, and balloon-borne instruments to gather data, meteorologists found that hurricane clouds might contain large quantities of supercooled water—the kind that had proven itself amenable to crystallization in other weather control experiments. And this supercooled water was found in "cloud chimneys," cumulus clouds with considerable vertical extent that ringed the hurricane's eye, allowing warm, moist air to flow up through the bottom of the hurricane and to be ejected out the top, thus serving as an energy cell keeping the hurricane "alive."[83] The working hypothesis was that this area was either inertially unstable or nearly so. If the hurricane were seeded near the eye, then the surface pressure would be modified due to additional latent heat release as moisture condensed around the seeds. That modification would then trigger inertial instability, the eye wall would migrate outward, and the hurricane's maximum winds would decrease due to the conservation of angular momentum—a spinning ice skater who extends her arms outward to slow down is illustrating this same concept. The ideal hurricane to experiment on would have a well-defined eye with winds that decreased with distance outside of the eyewall.[84] But before they could carry out the experiments, hurricane researchers needed a way to get the seeds to the target efficiently. The US Navy was prepared to help.

While NHRP personnel were trying to pry secrets out of hurricanes on the East Coast, out West personnel at the US Naval Ordnance Test Station (NOTS), China Lake, California, were developing an improved method of delivering silver iodide seeds to clouds, an outgrowth of their work to develop colored smokes for cloaking military operations. This new device used pyrotechnics, which allowed for the production and delivery of ice nuclei directly to the intended target, and was developed as part of Project Cyclops, which would later become Project Stormfury. According to the project director, geophysicist Pierre Saint-Amand, and his colleagues, they started "at the wrong end of the program" (i.e., with development and field trials, not with theory development) because many people were opposed to weather modification "in general," and some weather modifiers already thought they had solved the problem of delivering seeding material. Hence, they saw no need for a new technique, and there was a "widespread feeling within the scientific community" that only theoretical work was appropriate at the time: No theory? No practical work. Saint-Amand, by contrast, thought that "development and field work add to understanding and lend impetus to theoretical work," and hence he and his colleagues moved ahead.[85] A military weapons development center has considerably more leeway in moving to the development stage in the absence of firm theory if the scientists and engineers involved are confident they can create the desired outcome.

The first opportunity to test out the NOTS-developed pyrotechnic canisters (fig. 6.12) of silver iodide appeared several years later during Hurricane Esther. On September 16, 1961, a USWB aircraft dropped eight canisters into Esther's eyewall. As a result, Esther's barometric pressure, which had been dropping by one millibar/hour before seeding, stopped deepening and maintained a relative constant pressure thereafter. Seeding continued the next day, but the canisters fell wide of the eyewall, and the pressure remained the same. The experiment was deemed a success, and Stormfury became a joint Navy–Department of Commerce project in 1962.[86]

Further development of numerical weather prediction techniques also played an important role in Stormfury. Researchers, in particular Joanne Malkus (later Simpson) and Robert Simpson (of the Saffir-Simpson scale of hurricane strength), modeled the effects of silver iodide seeds on individual cumulus clouds, and from those results determined that "dynamic seeding," which sought to modify the motion of the air, was the appropriate approach for hurricane modification.[87]

Unfortunately, no suitable hurricanes appeared in 1962. In August 1963, Project Stormfury researchers randomly seeded eleven nonhurricane cumulus

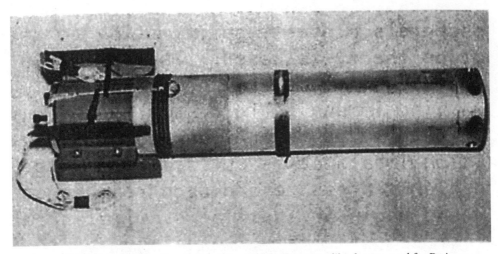

FIGURE 6.12. Navy Pyrotechnic Silver Iodide Generator like the one used for Project Stormfury. From National Hurricane Research Project Report No. 60.

clouds. Of the six seeded clouds, five reacted as predicted by the cloud model, and none of the five unseeded clouds behaved like the seeded clouds.[88] Buoyed by this success, the team was ready to take advantage of the opportunity presented by Hurricane Beulah, which developed just a few days later. Although Beulah was not an ideal candidate at first—Category One Beulah's winds were only ninety miles per hour and the eye was poorly formed—they seeded anyway, but ended up missing the towering cumulus surrounding the eye. They tried again the next day, by which time the winds had increased to 112 miles per hour (i.e., Category Three) and a distinct eye had formed. The seeding team managed to hit the target clouds. Airborne observers reported that the eyewall disintegrated, reformed with a larger diameter, and subsequently weakened. Wind measurements confirmed that the maximum wind speed had decreased by 20 percent and the zone of maximum winds had moved farther away from the center. Therefore, the results from Beulah matched those from Esther and those based on the original Stormfury hypothesis.[89]

From 1964 through 1968, Stormfury researchers were unable to seed any hurricanes because none passed within the "experimental area" set aside for the seeding experiment (see fig. 6.13). But the project continued. Researchers investigated alternative modification techniques, improved numerical models and pyrotechnic devices, and developed a new seeding hypothesis that involved seeding the first rain band outside the eyewall in an effort to cause a

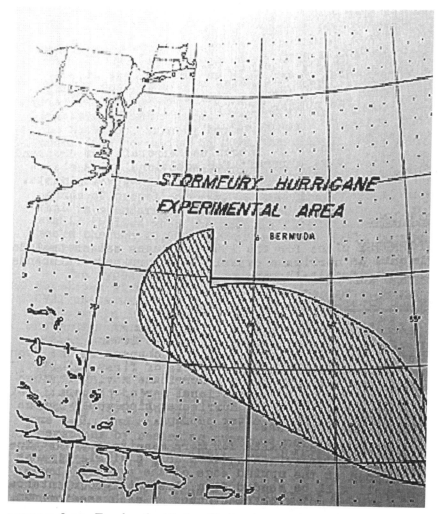

FIGURE 6.13. To reduce the risk of modified hurricanes striking land, Project Stormfury personnel could not seed any hurricanes unless they passed through this predetermined experimental area—a leftover precaution from Langmuir's 1947 modification attempt. From Project Stormfury, ESSA Fact Sheet, August 1966.

second outer eyewall to form, thus weakening the original eyewall. Once the eyewall reformed at the larger radius, the most rapidly moving air was further from the center and the storm weakened.[90]

 Hurricane Debbie in August 1969 provided Stormfury with its next opportunity to seed a hurricane, and researchers modified the technique by seeding

FIGURE 6.14. Interior of a Project Stormfury aircraft. Note the padding throughout the aircraft to prevent injuries during in-flight turbulence. NOAA Central Library.

the rain bands with massive amounts of silver iodide. Thirteen aircraft (see fig. 6.14) were involved in seeding and observing the results, and US Navy aircraft made five runs, dropping more than a thousand silver-iodide pyrotechnic devices each day. During the two seeding days, winds decreased 31 percent and 15 percent respectively, as predicted by models. After this apparent success, the Stormfury team was eager to seed another hurricane, but neither the 1970 nor the 1971 hurricane season produced suitable tropical systems. In 1971, the Stormfury scientists decided to seed Hurricane Ginger, but it was an ill-defined system and seeding had no effect. Then in 1972, the tropical storms either fell outside of the experimental zone or were too weak. A bigger problem: the Navy decided to pursue projects more in line with national defense needs and pulled its support and aircraft out of Stormfury. The USWB aircraft, which had seen ten years of hard service, were nearing their operational limits.[91]

At that point, the USWB looked to the Pacific Ocean in hopes of finding tropical systems it could seed, and it ordered two specially outfitted aircraft as observation platforms to join extant aircraft operated by NOAA, the Air Force, and NASA. However, the move to the Pacific was fraught with problems. It was one thing for the United States to seed hurricanes that might land on its territory, but another to seed hurricanes that might land elsewhere. If a seeded hurricane (typhoon in the western Pacific) strengthened and caused a huge amount of damage as well as loss of life in another nation, subsequent international outcry and legal consequences would have been huge. Once it became apparent that international agreements would not be forthcoming, it was impossible to move Stormfury to the western Pacific and Australia, and even in the eastern Pacific storms could strike Mexico or Central America.[92]

Even if Stormfury personnel had had a viable operational area and the necessary aircraft, the project could not continue unless other factors were met: improved navigational equipment and advanced instrumentation for recording meteorological data; the presence of sufficient supercooled water in hurricane convection cells to make seeding effective; a hurricane vortex that was sufficiently easy to change so that seeding would affect hurricane dynamics; and the opportunity to repeat the experiments enough times that seeding results could be distinguished from expected natural hurricane behavior. While the navigational equipment and the advanced instrumentation were present— indeed, they were the best available and turned out to have a significant impact on other atmospheric research—it turned out that hurricane convection cells contained too much natural ice and too little supercooled water. Researchers could not reliably distinguish the effects of seeding from natural behavior, thus eliminating statistical viability.

With the operational area in doubt, by mid-1983 it was apparent that a continuing program to reduce hurricane hazards was not a scientifically or politically viable undertaking. Thus Project Stormfury ended. While the objective of taming hurricanes went away, research missions to understand hurricane formation, structure, and dynamics, and to improve hurricane forecasts, continued.[93]

Stormfury's demise was not the end of the line for the Navy's pyrotechnic seed delivery systems. Indeed, Stormfury provided the perfect cover story for the military's efforts to develop a weather weapon—one that could be used to enhance monsoon rains in Vietnam and Laos and thereby bog down men and materiel making their way down the Ho Chi Minh trail into South Vietnam during the Vietnam War. The pyrotechnics were designed to operate on cumulus clouds, and they were tested on cumulus clouds in the tropics, which would be similar to those in Southeast Asia. Results from the Stormfury experiments provided Navy scientists with the feedback they needed to perfect their equipment and then turn it loose as a weapon of war, but first it would be tested as a state diplomatic tool.

CHAPTER 7

Weather Control as State Tool on Military and Diplomatic Fronts

We shall propose further cooperative effort between all the nations in weather prediction and eventually weather control.

PRESIDENT JOHN F. KENNEDY to the UN General Assembly, September 25, 1961

May the rainmakers succeed!

WALTER W. ROSTOW, President Lyndon B. Johnson's national security adviser, on the secret US/India weather control project, December 29, 1966

Even while using national security as a touchstone—it was the Cold War, after all—most political discussions about weather control focused on the domestic, "bring-home-the-water" variety. The State Department, however, was casting a wary eye on its international implications. Would it be possible, for example, to woo non-aligned nations with weather control without upsetting their neighbors? But agencies behind weather control activities—including the National Science Foundation and the Department of Defense—were not sinking money into investigations of these nagging questions. As they worked to fund basic- and applied-research projects, the concomitant problems that might accompany full-fledged weather control were left for the future.

At the same time, worries that Cold War adversaries might pursue the same goals prompted President Kennedy to tell the UN General Assembly that weather control should be pursued for the common good and not as a war-fighting tool. Public statements, however, do not always reflect classified research and its intended functions, and within a few years the United States was pursuing a weather control carrot for diplomacy and a weather control stick for military ventures. Both carrot and stick originated from the same classified cloud-seeding technique tested over southeast Laos in a research program dubbed Project Popeye. Project GROMET—a secret "agrometeorological survey" brokered with the Government of India by the State Department, but funded and executed by the Department of Defense—aimed to use weather control to simultaneously break a drought, keep nuclear weapons out of India, and bring non-aligned India into the US orbit. Its equally secret counterpart—Project Compatriot—used weather control as a weapon in Laos and Vietnam. Related to the unclassified domestic Project Stormfury, these projects provoked discussions and stoked worries at the highest levels of the federal government.

In 1961, the Naval Ordnance Test Station (NOTS) in China Lake, California, began working on weather control, an outgrowth of creating colored "smoke bombs" as high altitude markers. Over time, their advanced pyrotechnic delivery system allowed appropriate placement of silver iodide seeding devices in hurricanes and cumulus clouds. (See Stormfury, chapter 6). However, military weapons centers do not modify clouds just because they can. As NOTS geophysicist Pierre Saint-Amand testified during Senate hearings in 1965, "Primarily the work is aimed at giving the U.S. Navy and the other armed forces, if they should care to use it, the capability of modifying the environment to their own advantage, or to the disadvantage of the enemy. We regard the weather as a weapon. Anything one can use to get his way is a weapon and the weather is as good a one as any."[1] After discussing NOTS's collaborative efforts with civilian agencies, Saint-Amand remarked that NOTS was "about through" with basic R&D related to creating and delivering seeds for freezing clouds and was ready to begin field experiments for military purposes such as dispersing cold fog (warm fog being a much more intractable problem). It was time for large-scale experiments to determine the effectiveness of extant weather control techniques, and every agency involved with weather control research and applications should be pushing forward the agenda.[2] However, he did not want them interfering with military interests.

In light of future events, Saint-Amand's closing comments are interesting. He opined that international implications of "harvesting water" through cloud seeding had not been adequately addressed. At some point it would be possible to "control other people's climate" as well as America's. "Economic warfare," he said, would probably "come up in the future . . . and whether . . . it is intentional or not, we are going to be accused in the event we start taking a lot of water out of the cloud systems in Western Europe or perhaps in the United States."[3] The Naval Ordnance Test Station was not, according to Saint-Amand, actively taking steps to analyze possible social implications of weather control, but it was thinking about it. Whether he sensed it or not, within a year military, diplomatic, and social implications of weather control were going to become significantly more pressing.

A policy paper entitled "Foreign Policy Implications of Weather Modification" landed on Secretary of State Dean Rusk's desk in 1966. Threatening a nation's economy or security via weather control, intelligence analyst Howard Wiedemann wrote, was not yet possible, but it would be. Since the United States intended to conduct large-scale weather and climate control research, it needed to develop relevant policies before deliberately modifying weather that would cross an international border. He continued, "Further research may lead to opportunities for using weather modification techniques for common benefit, including technical assistance to less developed countries," or for inflicting massive damage on enemies. If one nation's weather control could adversely affect a downstream nation, then international control of weather and climate modification was necessary. In fiscal year 1965, the United States was spending $4.97 million ($35 million in 2015 dollars) on fog dispersion and rainmaking, and the Soviet Union was spending $20 million ($140 million in 2015 dollars) on all types of weather modification efforts. "In light of the steady increase in total Soviet R&D expenditures and its long-standing interest in weather control, it is likely," Wiedemann continued, "that the USSR will continue to devote substantial resources in this field."[4] Since weather control efforts were expected to expand in the next five to ten years, the American state had decisions to make.

Weather control presented opportunities and problems. Opportunities abounded for small-scale efforts: fog dissipation, modest rain enhancement, hail mitigation, and precipitation for hydroelectric power stations among them. Even with minor benefits, these small projects could help less-developed countries without adversely affecting adjoining nations. However, using weather control as a foreign aid tool could bite back. Nations might blame adverse environmental conditions on weather control despite a lack of causal connections, prompting diplomatic protests or worse. If the United States assisted a country via weather control techniques, it would be "essential to stress the limitations of weather modification in order to keep expectations within reasonable bounds; in collaborating with other countries on international projects, it may be difficult to strike a neat balance between healthy skepticism and an imaginative approach."[5] Strike a neat balance? The State Department's science desk learned just how complicated it could be when DoD offered to mitigate India's Bihar drought.

DANGLING THE WEATHER CONTROL CARROT: RAIN FOR INDIA'S WITHERED CROPS

India had gained independence from the United Kingdom in 1947—about the same time the Cold War grew colder. A poor, non-aligned country abutting southwestern Communist China and in the shadow of the looming Soviet Union to the north, India trod carefully between West and East. Jawaharlal Nehru, India's first prime minister, held that "the well-being of India and its poor masses depended on the adoption of modern technology . . . [relying] on science and technology, including the scientific spirit as valuable tools."[6] In 1947, that meant going after the biggest scientific and technological prize of all: nuclear power. Not only would mastering the atom lead to a prosperous country, it would "elevate Indian science onto the world stage."[7] The resulting state focus on heavy industry left less support for agriculture, and in a nation numbering 345 million mostly impoverished people in 1947, food would be a continuing problem. By 1960, the population had grown to 450 million, and the United States was providing significant amounts of food aid to India. But regardless of food support, India remained non-aligned, even as its rival Pakistan entered into regional treaties with the United States, exploiting the latter's anticommunist policy—a relationship that led Indian leaders to distrust US diplomatic efforts.[8]

It is against this backdrop that we consider the geopolitical situation of the 1960s and how the United States came to offer weather control to the Government of India as a carrot to shift its sights to the West. Between John F. Kennedy's inauguration in January 1961 and DoD's offer to put weather control to work against India's Bihar drought in late 1966, India came out on the losing end of two major geopolitical events that diminished its global standing: the 1962 Sino-Indian war over the disputed Aksai Chin plateau, and the 1965 Indo-Pakistan war over the disputed Rann of Kutch and Kashmir regions.[9] But India's national pride was especially stung by China's explosion of a nuclear device in 1964, effectively usurping India's claim to the scientific and technological leadership it needed to maintain influence with Asian and African nations. For India, building a nuclear weapon seemed like the logical way to get back in the game, and pressure to do so began to mount within the country.[10] But the United States, already unhappy about China's entry into the nuclear club, did not want to see any more members, not even a non-aligned one. Seeing nuclear weapons as a security issue—not as a way for a nation to increase its status in the eyes of its neighbors—meant that the United States was looking for a way to help India achieve its science and technology goals without building the bomb.[11]

In the midst of these border disputes, Nehru died and was succeeded by Lal Behadur Shastri, who in turn died in January 1966, just hours after the Soviet-brokered peace agreement between India and Pakistan was signed in Tashkent. Later that month, Nehru's daughter, Indira Gandhi, became prime minister, inheriting a huge deficit, a substantial balance-of-trade problem, and a worsening food shortage exacerbated by a deepening drought that had started in summer 1965 with the failure of the summer monsoon—the rainy season that provides moisture for India's grain-producing farmlands.[12] Just two months later, Gandhi made a state visit to the United States, seeking help from President Lyndon Baines Johnson.

United States presidents do not usually get involved with the nitty-gritty day-to-day food aid work of the US Agency for International Development (USAID). But Johnson, desirous of spreading his Great Society throughout the world, did not fit the typical mold. Johnson trusted no one at the State Department other than Secretary Rusk, so he decided to use Public Law 480 (the Agricultural Trade Development and Assistance Act of 1954—later called "Food for Peace") to provide food aid to India and several other countries. The advantage to Johnson: USDA, which managed Food for Peace, was led by Johnson ally Orville Freeman, and PL-480 allowed for short-term aid decisions while other aid programs were strictly long term.[13] Essentially, Johnson became the "desk officer," personally distributing food aid to India.[14] But the 1965 drought significantly worsened the food problem, so Johnson ramped up his efforts. Not content with addressing the immediate need to feed millions of people, he used his prodigious political savvy to exploit the Bihar drought in pursuit of his ultimate goal: making India self-sufficient in food. And that meant forcing the Indian government not only to make agriculture an economic priority but to undertake fundamental agricultural reforms that it had opposed for years.[15]

As the drought deepened in early summer 1965, Johnson executed his "short-tether" policy, that is, he released grain so it would arrive "just in time." Each month he analyzed the situation, and then he, and he alone, decided whether to permit grain shipments. No grain left the United States for India without Johnson's express permission. As one might expect, Indian leaders were not only unnerved by not knowing whether they would get grain for their people, they resented Johnson's overbearing attitude and tactics. Yet Johnson's decision had the desired effect: the government significantly increased its investment in agriculture, and in the hands of its reform-minded agriculture minister, Chidambaram Subramaniam, India was finally making progress that boded well for the nation's food production.[16]

When Prime Minister Gandhi visited Johnson in March 1966, she pressed India's case for more food aid, stressing its serious effort to reform agriculture as it moved toward self-sufficiency in food. More work needed to be done, and she wanted Johnson's assurance that India would get the food it needed to stave off mass starvation. Johnson reassured Gandhi . . . sort of. If Congress were willing to support food aid to India, it would continue, but they discussed and agreed that other nations needed to step in and provide food assistance too.[17] India's food production depended on water, and absent irrigation, water came from rain. No amount of agricultural reform could make up for failed summer monsoons. As the grain harvest plummeted once again, Secretary of Agriculture Freeman told Johnson that India's situation would be desperate by fall 1966. He wanted more fertilizer shipped to India. "The weather for next year's crop cannot be controlled," he wrote, "but the amount of fertilizer to be used can be."[18] Freeman was right about the fertilizer, but not about the weather, because attempts to control it were already in the works.

As Johnson kept a close eye on India's agricultural output and its weather forecasts—he later recalled that he knew "exactly where the rain fell and where it failed to fall in India"—State Department personnel were discussing the Wiedemann paper on weather control. They recommended to Johnson that State's India desk consider options to keep India from "going nuclear" as they took steps "to enhance India's political prestige, including scientific and technical projects." Johnson approved.[19] If "dramatic uses of technology to attack India's basic problems of food, population, health, and education" could be forthcoming, that would be even better.[20] As State's science desk noted, India had long-running scientific strengths in meteorology that might lead to a cooperative effort to place a geostationary satellite in orbit over the Indian Ocean, significantly helping with monsoon forecasting.[21]

But in late 1966, Johnson—still manning the food aid desk—quit shipping grain to India, much to the consternation of Americans and those abroad who were anticipating mass starvation in India. Ignoring criticism that rained down from all sides, Johnson, determined to force Gandhi to complete agricultural reforms, refused to budge.[22] The continued drought, however, was the main impediment to advancing the agriculture system. India needed water, and it needed it soon. Coincidentally, the State Department was still looking for a scientific and/or technological project that could raise India's national self-esteem at the same time that it solved a state problem. The Department of Defense also had a problem in need of a solution: it needed a new venue to test the latest NOTS-developed pyrotechnic silver iodide delivery system. Within a few months, these needs and wants converged to the same solution: weather control.

PROJECT POPEYE: TESTING A WEATHER WEAPON IN LAOS

In early August 1966, the chief of naval operations contacted NOTS about an experimental program to be undertaken in Southeast Asia. Its mission: use the NOTS-developed silver iodide delivery system to increase rain and then evaluate the results, with effectiveness based not on how much rain hit the ground, but how many Viet Cong infiltration routes into South Vietnam were muddied up and made impassible. The selected test area was part of the Annam Mountain range, which lay mostly in Laos, with some straddling of Laos and Vietnam. Mountain elevations extended as high as 7,500 feet, so it was a good place to seed orographic-enhanced clouds, and it fell within the monsoon belt, which meant a dry season from November to March, and a wet season from April to October. Dubbed Popeye, the program's goals were to enhance rainfall during the main part of the wet monsoon and to extend its starting and ending dates.[23]

From the time the NOTS team had started working on weather control techniques in the early 1960s, it had gained significant data about tropical cumulus cloud seeding during Project Stormfury, and about orographically enhanced cumulus cloud seeding during the Sierra Project, carried out in California's Sierra Nevada Mountains. The NOTS team had also developed several types of nucleation devices that could be used on the ground, mounted on aircraft, or carried on an aircraft and then targeted on a cloud some distance away. Popeye team members used four different types of silver iodide pyrotechnic generators (all suitably named to maintain the Popeye theme):

- Sweetpea, a small flare producing 110 grams of silver iodide smoke,
- Wimpy, differing from Sweetpea only in the way it was fired,
- Grumper, a large aircraft-mounted flare producing 230 grams of silver iodide smoke, and
- Bluto, containing a special propellant composition burned in an aircraft-mounted rocket motor and producing 940 grams of silver iodide smoke.[24]

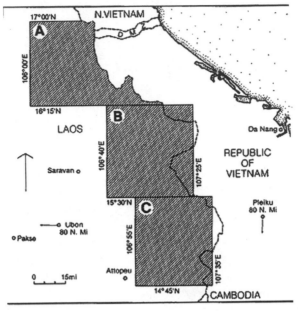

FIGURE 7.1. Project Popeye test areas. From E. M. Frisby, "Weather Modification in Southeast Asia, 1966-1972," *Journal of Weather Modification* 14 (1982): 1–6.

The Popeye experiments (fig. 7.1) took place in fall 1966, with the test team using a variety of military aircraft for seeding and observations. Other military personnel, including weather observers, who came from Navy, Marine, and Air Force units, joined pilots and air crews.[25]

Gathering experimental data despite operating under occasional combat conditions, the team members supplemented field data with surface and upper-air weather maps covering Vietnam and Southeast Asia, satellite and radarscope images, and moving and still photographs. They also maintained detailed logs of events inside and outside the aircraft.[26]

Quantitative results were lacking, but the combined qualitative data indicated that seeded clouds grew six to ten times taller and wider within ten minutes of seeding and doubled the precipitation of unseeded clouds. This success led the commander in chief of the Pacific region to conclude that "cloud seeding to induce additional rain over infiltration routes in Laos could be used as a reliable tactical weapon."[27] The Department of Defense wanted to make the project operational, but the wet monsoon season wouldn't return until April and seeding couldn't start until March. Where else might it pursue this technique? India's Bihar drought provided the answer.

PROJECT GROMET

In early December 1966, Secretary of Defense Robert McNamara sent an EXCLUSIVE FOR cable to the US ambassador to India, Chester Bowles, who was interested in DoD's offer of a "Joint U.S.-India Precipitation Experiment" using "new technology to increase the rainfall in India." McNamara cautioned Bowles against being too enthusiastic. Defense had only tested the technology in one (unnamed) area under special conditions, and until researchers knew it worked under India's unique environmental conditions, it was unwise to lead Government of India (GOI) officials into thinking it was a sure thing. The best results would come in May with the summer monsoon, but DoD personnel were willing to start seeding in January over drought-stricken Bihar and Uttar Pradesh "in an attempt to improve crop yields." (McNamara neglected to say that May would be outside the test window—seeding in Laos would start in March, so DoD had no reason to wait until May.) Caveats explained, McNamara urged Bowles to meet soon with Prime Minister Gandhi to discuss this possibility. A technical adviser could be in Delhi immediately if Bowles needed help making his case. If Gandhi agreed, DoD would schedule a meeting between Indian and US scientists and develop a plan; a three-man team could be in Delhi with a "few days notice" and start the operation in three to four weeks. The DoD assumed India would bite, and it was arranging for aircrews and equipment.[28]

With the technical bits addressed, McNamara continued on to the "international, legal, and political problems" inherent in weather control projects. He emphasized that the project needed to remain "entirely covert and with the understanding, preferably in writing, that India will hold the United States harmless against any liability for damage that may arise from this program." Anticipating success, DoD would work to make the technology available to civilian agencies for a "publicized joint program with Indians." To provide additional cover, McNamara thought DoD might bring in an outside weather control expert from Australia. Defense and State were in accord. Bowles needed to get back to him on whether the covert nature of the program would be acceptable.[29]

Days later, State Department personnel were discussing a memo entitled "Study of International Law and Politics of Weather Modification," which mentioned that a preliminary study needed to be done, preferably within six months. They did not have six months; cloud seeding in India was moving ahead rapidly. Apparently the memo's drafter had no idea what was being proposed for India, and DoD's representative concurred there was insufficient lead time for a study.[30] Even before the project started, State was wondering if it should extend the project into Pakistan to head off any international incidents over India "stealing" Pakistan's water. Defense did not think so—there would not be enough cloud cover.

Some NOTS pyrotechnic devices had been sent outside of the country previously, when Saint-Amand shipped twenty devices to the University of Jerusalem's Department of Meteorology for testing without notifying the Israeli government. He had provided the unclassified Stormfury devices, not the "smaller, more sophisticated" classified models destined for India. State also learned about another "project of interest" to both State and Defense—likely Popeye or Compatriot—but the DoD representative was emphatic that these programs not be lumped together.[31] Tying GROMET to Popeye would have been a security disaster for the administration—a chance it could not take.

Herman Pollack, of State's science section, advised Bowles that State was not as gung-ho as its Defense counterparts. The seeding technique was secret, the project was military-based, January was not an ideal seeding month, there was a very real possibility of raising false hopes among the Indians in the drought zone (even DoD was a tad uneasy on this point), and conducting an experiment that might affect weather on the other side of already contested national borders could create international legal nightmares. State hoped that the DoD briefers' enthusiasm was not misplaced. Pollack and his colleagues were eagerly awaiting the results of the first experimental runs.[32]

The loop among all of the players started to close in late December, when Presidential Science Advisor Donald Hornig informed National Security Advisor Walt Rostow that Project Popeye had been successfully concluded, and DoD was ready to go operational. However, President Johnson needed to approve the plan.[33] Following the Christmas holiday, Rostow reminded Johnson of the planned "highly classified" experiment to "seed clouds over Bihar and Uttar Pradesh in an effort to make some rain available within the next two months. If we succeed, this will materially improve the chances that this spring's crop will produce something in the worst affected areas." To maintain a low profile, participants would fly in small commercial planes and occupants would wear civilian clothes—whether civilians or not. If the media started asking questions, they would respond from a mutually agreed-upon press release that would call the project an "agro-meteorological survey." Rostow finished with a flourish: "May the rain makers succeed!"[34]

The same day, State and DoD notified Bowles that the United States and the Government of India (GOI) had agreed to participate in GROMET, which, they stressed, was a classified program and all correspondence related thereto would be EXDIS (Exclusive Distribution), that is, it was extremely sensitive and restricted to high-level personnel. Under the terms of the memo of understanding, the project would take place in Bihar, eastern Uttar Pradesh, and other areas designated by the GOI. Once the experiments were over, personnel from both countries would collaborate on evaluating the results. While the GROMET team prepared to take flight, Bowles continued to work out administrative details with the GOI. The United States would provide personnel (federal employees or contractors), aircraft, instruments, and seeding materials. The GOI would provide aircraft consumables (fuel, lubricants) and arrange for room and board. It would also hold the United States harmless for any injuries or damage. Both nations would agree on public press releases.[35] What the GOI declined to do was to waive "red-tape clearance procedures." Once the State Department made clear it was "disappointed," however, the GOI suggested that all participants enter the country as tourists.[36] Problem solved.

While Bowles worked out the diplomatic kinks, the NOTS team was preparing for the trip to India. At the top of the list: finding pilots for the seeding aircraft that were to be airlifted to Delhi. One of those contacted was George A. Brown. Brown had no idea what he was getting into when "Lee," brother of his friend "Howard," called and asked if he had any plans for the next few months. Intrigued, Brown called on both of them. Lee (there are no last names in this account), who worked at NOTS, asked Brown if he could get a passport so he could "leave the country for a few months." Lee had already thoughtfully arranged for a secret clearance via the State Department and could only say that the project was called "GROMMET" (with an extra "M"—probably due to a mispronunciation). Brown was given the rank of Air Force lieutenant colonel of noncombative forces, Far East Asia, plus a substantial check. Brown also discovered that he was to meet an Air Force C-141 Starlifter at China Lake in four days' time. Once Brown agreed, Howard pulled out an envelope labeled "SECRET" and handed it to him. While Brown read the enclosed document, they drank twenty-year-old cognac and smoked cigars. The plan—an absolutely secret plan, they reminded Brown—was that they would load four turbo Cessnas into the belly of the C-141, fly them to India, and conduct rainmaking experiments.[37]

After lifting off from China Lake, Brown, Lee, Saint-Amand, and the rest of the team made a stop at Travis Air Force Base in Northern California for a crew change before heading west on the thirty-hour flight to Delhi—smoking lots of cigars en route. In the early morning of January 23, the Starlifter descended from the dark, cloudless sky to land at Palem Airport outside Delhi. Project GROMET had come to India.[38]

Over the next few days, Brown and the aircrews reassembled the four Cessnas while Saint-Amand and the scientific team prepared for seeding. Parked next to the Cessnas was an Indian Air Force squadron composed of Soviet MiG-21 fighter jets. On the other side, an old DC-4 perched on its collapsed landing gear, its wing providing the team shelter from the hot sun. What was in the fields surrounding the landing strip? Snakes and cobras—the team called them "one, two, and three step snakes" for how far one could walk after being bitten—that made their living catching rodents and other small wildlife.[39] Saint-Amand met with Indian officials from the Agriculture Ministry, the meteorological service, and civil aviation to schedule scientific meetings and arrange for appropriate communications frequencies and ground control for the flights.[40]

A far thornier problem than settling ground arrangements was agreeing on responses to possible press queries. A major sticking point: how to respond to questions about US military involvement. For the GOI, the answer had to include "no," as in, "no, the US military is not involved." For State, the answer needed to be more nuanced: "This is an agriculture experiment. Next question, please." Not a flat denial, which would be a lie, but limited information. The same problem pertained to secrecy: "It is not being kept secret—witness the fact that we are answering your questions," not "What project? Who said we had a project going on here?" Address the question—give almost no information. Another major challenge: the election campaign season was underway in India, and the approving officials were not in Delhi. The two weeks of diplomatic efforts must have seemed an eternity to Bowles, who needed an agreement before the first aircraft, and silver iodide seeds, flew.[41]

By the time everyone had signed off, some GOI officials were wondering if it made more sense to announce what they were doing. However, Bowles and his primary contacts within the GOI thought it a bad idea to make any public statements until they had had a successful seeding run. And GROMET team leaders had suggested that they use the technique outside of the previously agreed-upon experimental area to ensure they had a broader range of successes before making any statements. Bowles concurred.[42] The final agreed-upon contingency statement:

Scientists from the United States and India are cooperating in a joint agro-meteorological research project, localized in the Eastern Uttar Pradesh and

Bihar (and any other area to be indicated by the Government of India) to study the cloud physics and rain producing mechanisms over these areas of India which have incurred several droughts over the last few years.

According to these negotiated answers, the research involved measuring standard atmospheric variables (i.e., moisture, air temperature, cloud top temperatures, etc.) and experimenting with cloud seeding. If the query concerned India's past experiments with weather modification, they were to note that India had been working on "artificial stimulation of warm clouds" in northern India, and this current experiment would be on cold clouds. Not said was that Indian scientists had been working on weather modification techniques since the 1950s. In case people wondered why Americans were involved if the Indians had had their own projects in the past, they were supposed to be told that previous cloud-seeding attempts had all been ground-based, while this joint project involved "very high altitudes with especially equipped aircraft." Were they making rain? They were experimenting with "stimulating rain" in a small area, but it was important to remember that conditions were not conducive during this season. Has this worked before? Promising results had come from experiments conducted in the Caribbean.

Subsequent questions were a little more pointed. *Question*: "Is this an experiment which the Americans are making for their own benefit and in order to perfect techniques which are considered to be too risky for use in their own country?" *Answer*: We've tried this in America, we don't think there is any risk, and it would be a good idea to test it where rain is needed. *Question*: "Could this technique be used clandestinely to cause inordinate rainfall downstream, thus flooding the territory of any other country?" This question, as initially posed, suggested that the country would be Pakistan. Cooler heads decided that even if that were the question posed, they did not need to anticipate quite that level of detail in their printed questions and answers, in case a copy fell into the wrong hands. The answer: "Effects of this technique are strictly localized and of short duration." Another great question: Has the government guaranteed that the people in these planes won't be spying? Not a problem. Each plane would carry a member of the Indian Air Force, and Indian scientists were involved with the project.[43]

By early February, the planes were ready for a test run to check out their equipment. According to pilot George Brown, their mission was to fly three aircraft north of Delhi, climbing to twenty-six thousand feet with payloads of silver iodide rockets and flares. As they flew out of Delhi, they discovered the major flight hazard they would encounter: vultures and hawks that would take down their planes if they hit them during climb-out. The pilots decided to keep their airspeed down until they had cleared the large birds, and they also kept a close eye on their navigation equipment lest they stray too close to Pakistani or Chinese airspace, particularly since a small Pakistani plane had been shot down a week earlier when it strayed over a military zone. Sighting their first clouds, they fired their rockets and watched as silver iodide saturated the cumulus clouds, which started to rain. Then the flight got a little more exciting. As Brown's plane entered another cloudy area, Brown activated the rocket under the left wing, which promptly exploded. Caving in the bottom of the wing under the fuel tank, it caused the plane to start a roll. The blast also shattered a window, allowing extremely cold high-altitude air to flood the plane. Brown warned others not to fire their rockets, and they limped back to base. After modifying the rockets so they were safe for future use, the team members told their State Department handlers that they were ready to start the GROMET experiments.[44]

The team was ready, but nary a cloud appeared in the critical Bihar area. Meteorologists had reported, however, that clouds were forming in the northern Punjab. In GROMET's version of "Mother may I?" Bowles inquired: Can we seed there? Because of the recent shoot-down, State and Defense were less than enthusiastic.[45] Bowles realized the problems inherent in seeding near the Pakistani border—they were just trying to find some clouds to seed.[46] Since the GROMET project was "sensitive," State wanted to control seeding outside of Bihar and Uttar Pradesh. The GROMET team could suggest alternate seeding areas, but it had to float those ideas to State before raising them with Indian nationals. If GOI officials suggested a target area, State wanted to hear about it first.[47] But Bowles persisted. He had discussed the matter with Prime Minister Gandhi's secretary Lakshmi Kant Jha, who agreed that GROMET needed to take advantage of every opportunity "that [did] not entail risk of [an] international incident." If possible, it was extremely important "to demonstrate . . . that India's food and agriculture need not be entirely at the mercy of weather vagaries." Any success would provide India with a psychological boost and a practical agricultural benefit. They had invested a lot of money in this project because Bihar and Uttar Pradesh needed rain, but they could not seed clouds that were not there. Since the clouds were not coming to them, they needed to go to the clouds. And that meant they needed the flexibility to move when the opportunity arose without having to get permission from State. A dispirited Bowles wondered why the "sensitivity [was] so great" that their local judgment was in question. He also pointed out the obvious: they would not reap any benefits from the project if they could not publicize an ultimate success.[48] Less than twenty-four hours later, Bowles sent a second message to State—this one "immediate"—which State then forwarded on to the White House. Clouds were finally moving eastward across central Rajasthan. Could GROMET move to this site for seeding operations? Answer: No, for the present. The decision disappointed Bowles's contact at the Agriculture Ministry, J. C. Mathur. He was embarrassed because he had gone out on a limb with the Indian Ministry of Defense, convincing them to give the GROMET team permission to fly outside Bihar-Uttar Pradesh. "I hope," a dejected Bowles wrote to State, "you can give us . . . flexibility soonest."[49]

After "further consultations [with] Washington agency experts," State relented. The GROMET team was sufficiently competent to make judgment calls on downwind effects that might create potential international problems. However, State still insisted on being informed of any changes in area, and any seeding it performed had to have "some legitimate agricultural use." The GROMET team was "not repeat not" in India to demonstrate the technique, which might have been a surprise to DoD. Furthermore, there was to be no publicity until civilian agencies had taken over the project. State felt that "premature disclosure could kindle interest elsewhere which we are not prepared to consider."[50]

Given the connection to Project Popeye, State did not want interest "kindled" anywhere. It wanted the GROMET team to get in, see if it could make rain to benefit agriculture, then get out with as few questions as possible. Defense had wanted a test site, but not at the expense of blowing its cover in Vietnam and Laos.

The GROMET team finally located seedable clouds on February 12 and continued seeding over several days. These runs left one Cessna with a hail-damaged wing. Another flew too close to the border with China, which Brown discovered when "20 millimeter cannon slugs" from a Chinese MiG-21 "arced in front" of his Cessna. The next day they made sure to stay well south of the border.[51] Resulting rain amounts were mixed, ranging from heavy to moderate to light depending on the size and types of clouds seeded. In general, larger clouds produced more rain.[52] But then again, larger clouds contain more moisture, so that was not exactly unexpected. A situation report sent in late February provided similar results, with clouds responding as expected depending on their size and type.

In his summary, Bowles indicated that seeded clouds rained when they probably would not have naturally, but the dry air evaporated the moisture before it hit the ground. Even in areas where it appeared the rain reached the surface, they had no way of measuring rain totals—an inherent difficulty in a project that was supposed to support agriculture. Bowles noted that the GROMET team believed "that economically valuable amounts of rain can be produced over much of India during and after the monsoon season when there is usually abundant non-raining cloud cover."[53] Since the only moisture that matters to crops is what lands on them, it was difficult to call the project "successful." And given reports that knowledge of GROMET was "spreading in India," the GOI and Bowles had to figure out what to say. J. C. Mathur wanted to release a public statement soon since Parliament would convene in mid-March, and questions could arise then. Bowles was confused by State's directive to release no statements until the project had been taken over by civilian agencies. If civilian agencies did not take control until after the experiments were "successful," State might be waiting "a long time, indeed longer than is politically prudent at this end." Therefore, Bowles wanted State's permission to meet with Mathur and work out a joint statement indicating the project had been "useful and both sides [were] giving further study to possibility of continuing it." In the meantime, Bowles and his team in New Delhi were trying to figure out how the project might be continued, but he stressed that State's decision on a "low key public statement" should not depend on his draft plan.[54]

When Rostow reported GROMET results to President Johnson at the end of February, he ended, "State and the scientists are sorting out what kind of a statement to issue—if any."[55] The archives yielded no additional messages between Bowles and State on a "low key public statement," so it seems that State's answer was "no." In addition, two relevant books written by Indian authors in the late 1970s do not mention government rainmaking efforts in India during this period, casting doubt on supposed "widespread" knowledge about GROMET.[56]

Despite the lack of success, State wanted to extend GROMET to Pakistan. Writing to Bowles in New Delhi and Ambassador Walter P. McConaughy in Rawalpindi, State declared that it had undertaken an eight-week project in India due to urgent humanitarian needs related to the drought, and discussed the limited nature of the cloud seeding and the west-to-east winds that limited seeding's impact to Indian territory. With the summer monsoon starting in April, winds would blow from east to west, which meant seeding in India could impact Pakistan, particularly if clouds that would have rained out in Pakistan rained out in India instead. State did not want Pakistan to conclude that India was stealing its water, a very real possibility since Pakistan, as McConaughy noted, had an "almost psychotic fear of India."[57] Pakistan needed water too, so politically it made sense to cover both countries. State wanted to discuss the possibilities with the GOI and then with the Government of

Pakistan. May would bring "good cloud hunting," and seeding teams needed to get started.[58]

Nothing, apparently, ever came of this proposal, although State considered sending scientists to Rawalpindi to discuss the matter with the Government of Pakistan.[59] Indeed, the idea of using weather control in India started falling apart in spring 1967. Writing to Science Advisor Donald Hornig in May, Bowles expressed worries because he had received no "green light" to extend GROMET and turn it over to a civilian agency. He thought seeding would continue into the summer monsoon season, which is how he had pitched the project to Prime Minister Gandhi and why he had expended so much time coordinating GROMET with State, Defense, and the GOI. Bowles understood the problems involved with getting long-term programs underway. However, "right now," he wrote, "the hour for Indian democracy is late." India needed a significantly better harvest or the "fragile Indian democracy" could fall apart as its increasingly desperate citizens competed for food. The monsoon clouds were arriving. Could Hornig help?[60]

Bowles may never have known that US/India foreign policy was not responsible for GROMET's failure to become part of a civilian agency. A letter from Hornig to Johnson in early June explained that a US Agency for International Development (USAID)–sponsored "team of experts" was heading to India to begin some "*cloud seeding experiments* using a new technique." Because of weather control's promise, a special report issued by the Federal Council for Science and Technology advocated for a tenfold or more increase in funding for the Environmental Science Services Administration and Interior as they pursued advanced weather control techniques. However, Hornig warned LBJ of serious political implications involved with using weather control techniques.[61] Hornig's statement about weather control in India is not supported by any other evidence. The GROMET experiment ended, and while State had suggested that three scientists might visit both India and Pakistan to discuss further weather modification efforts, nothing apparently came of it.

Even without GROMET, the summer monsoons provided plenty of water to India's parched land; when this water combined with improved seeds and additional fertilizer, India harvested a bumper crop of grain in 1967. As the threat of famine faded, State did not need cloud seeding to assure a better harvest, and Defense did not need a connection made between strange-looking modified clouds in India with even larger, strange-looking modified clouds in Vietnam and Laos. The experimental Project Popeye was about to turn into the operational Project Compatriot, as the United States turned weather into a weapon.

DEPLOYING THE WEATHER WEAPON: VIETNAM
AND LAOS

As the GROMET team hunted for suitable clouds in India, Hornig was updating President Johnson on Project Popeye. He had previously discussed the project with Johnson—its purpose was to muddy up the trails that brought men and materiel into South Vietnam—and cautioned him that this new cloud-seeding technique held considerable military potential, but could present "serious political problems."[62]

The seeding technique proved effective in Vietnam and Laos because of large cumulus clouds present before and after the summer monsoon. When the pyrotechnic canisters full of silver iodide hit the clouds, the latter "blew up" into a distinctive shape, the same shape that observers saw in India, and one reason GROMET ended. The Popeye tests had induced significant amounts of rain from 80 percent of the seeded clouds and had intensified some storms. Hornig reminded Johnson that even though the technique had been successful, whether it would reduce the supply flow or have "adverse side effects such as flooding populated areas or crops," was unknown. The question: should the United States "initiate a full-scale operation" or not? The Joint Chiefs of Staff supported an extensive plan that would cover the length of the Laotian Panhandle and southern North Vietnam, and they wanted to get started "as soon as possible." Hornig wrote that State was not opposed, but was concerned civilian areas could be damaged and crops destroyed by seeding too close to coastal southern North Vietnam. Consequently, State wanted to see more information on military effectiveness and was recommending an experimental run focused on "choke points."[63]

Hornig was troubled by the political implications of "weather warfare," noting that if the weather weapon went operational, the public would find out eventually, as would America's allies and enemies. "The charge will be made," Hornig wrote, "that we have been the first to open a 'Pandora's box' of unknown threats to future civilizations." Referring to some of the more controversial ideas for modifying the weather and how they could be used to deter enemies, Hornig noted that the Popeye techniques could create "disastrous floods in highly populated and intensive agriculture areas in certain parts of the world."[64]

Despite the potential downsides, Johnson gave the go-ahead, and Project Compatriot started on March 20, 1967. As an operational project, Compatriot was less cumbersome to military personnel than the experimental Popeye. They no longer had to maintain extensive documentation or randomize clouds. The pilots determined the best location to release the pyrotechnic devices based on air temperature and updraft. Three WC-130 "Hercules" weather reconnaissance aircraft and two RF-4C "Phantom II" reconnaissance aircraft were devoted to Compatriot; the former carried 104 seeding units on both sides of the aircraft, the latter 104 in its photo cartridge compartments. In addition to the four different types of seeding units used in the Popeye experiments (discussed above), personnel in China Lake continued to modify the seeding delivery units for a variety of cloud types and flying conditions, deploying all of them during the war.[65]

Two months later, Walt Rostow repeated Hornig's earlier concerns when he wrote Johnson a "TOP SECRET—LITERALLY EYES ONLY" memo about a suggested expansion of Compatriot that would flood the upper Song Ca watershed and subsequently cause landslides, hamper the operations of locks on coastal waterways, force supply movement to higher ground, generate twelve to thirty-six inches of coastal flooding from late July through early October, cause flash flooding in the highlands, and silt up navigable waterways, thus providing a significant military advantage to American and South Vietnamese forces. Despite what appeared to be serious consequences for those living in the area, a "careful intelligence analysis" determined there would be "no significant danger to life, health or sanitation" because the local population lived in homes on stilts. Seeding would destroy about 10 percent of North Vietnam's rice production, requiring Hanoi to increase imports by 70 percent, most coming from Communist China. "Psychologically and in terms of possible exposure of the operation," the report continued, "it raises a significant question of civilian damage in relation to military impact."[66]

It also raised questions about how the plan might be exposed. Chances of a leak were considered minimal—very few people knew about the project, and suitable cover stories were in place. However, if the North Vietnamese started accusing the United States of triggering the heavy rains and others governments started repeating the accusation, that could be a serious problem. The results of GROMET would be published in about a year, and it would take little effort to connect making rain in India to flooding Laos. If the North Vietnamese triggered speculation that metastasized into a widespread controversy, or a single well-regarded meteorologist started asking about possible military uses of weather as a weapon, the US government would have to provide answers. Therefore, the Johnson administration needed a plan for responding to queries. Rostow considered two options. They could outright deny using weather as a weapon and then continue to use it, counting on security measures to keep it under wraps. That might work for about year, he thought, but not much longer. If someone "leaked" information about the operation, the administration's credibility would be shot. Or, they could admit they were using weather as a weapon and then continue using it if they justified its military importance. Rostow thought the second choice was "least bad."[67]

Rostow then turned to "basic policy considerations." They needed to consider the public's conception of "weather warfare," which was no longer dropping dry ice to clear runway fog, as had been done in Korea, but had morphed into "climate or broad scale weather control," aided and abetted by newspaper articles. The US government had already taken "a beating" over "bombing, chemical agents, and even napalm," made worse because "we are a big and sophisticated nation making war on a small and backward one." Weather control, particularly if it flooded a substantial portion of the country, was not likely to be viewed favorably. However, Rostow continued, "objectively, a strong argument can be made that the Song Ca operation is essentially a form of interdiction, and as such much more humane than bombing"—an interesting comment since many consider flooding among the worst natural disasters because it does not go away quickly. During a drought, water may not be plentiful, but it is usually still available. And drought does not make travel next-to-impossible, spread disease, or destroy property. Floods do all of these things. As to crop damage, Rostow wrote that they could just argue that the North Vietnamese had damaged 10 percent of the South Vietnamese rice crop—so essentially the nations would be even. And they did not anticipate any "significant" loss of life or health impact, which could be checked "so that we would know at all times what we were doing."[68]

Rostow also bemoaned the fact that such "objective arguments" would not "overcome emotion." The United States could be accused of "using science unfairly against a small nation," and of using rainfall to flood North Vietnam even though it had promised not to bomb dikes. The North Vietnamese could distort, perhaps "grossly," loss of life and the extent of damage, leading the world community to pressure the United States to stop using weather control techniques in the war effort. "In either event," Rostow wrote, "our only defense would be the logic of our arguments based on a candid admission that rain making was being used as a humane and effective non-lethal supplement to our bombing of the North to reduce the flow of men and supplies into the South." The other pitfalls? If the United States claimed that it only used weather control in situations where it would not harm civilian life and health, would military operations in the future be held to such a standard? What if the United States were in a "major conflict"? "Would we, for example," Rostow wondered, "have rejected partial flooding of the Netherlands if this could have got us to Berlin two months sooner?" Perhaps the solution was to deal with the situation if it arose, and be more concerned about the opponent than international lawyers.[69]

Wrapping up, Rostow maintained that his analysis was "as accurate and thorough as possible." To get the "major military effect," the administration would risk compromising America's "image as a responsible and moderate nation." They could just continue with the original plan and induce localized heavy rainfall in the Laotian mountains where there was "no significant civilian damage whatever." This would be a "half a loaf" and could still offer the possibility of having to defend it "against illogical argument and accusations." But going forward with the Song Ca operation would be more likely to expose the Laotian seeding as well, and then they could lose the entire opportunity if they had to stop weather control. Considering the Song Ca operation on its own, Rostow noted that they did have "options short of the full theoretical 6.7 inches per month of additional rainfall, continued right through the rainy season." (It is nothing short of amazing that the estimate was down to a tenth of an inch.) They could call off the seeding at any time, especially if the North Vietnamese were catching on and ready to make public charges about the excess water. But a "limited and controlled" operation would produce a smaller military payoff, and the United States might be accused of using weather control anyway. Rostow's conclusion: add the Song Ca operation and "stick to it no matter what." Since the meteorological conditions were favorable, the seeding teams needed to move forward immediately, or wait for a year.[70]

Rostow requested Hornig's opinion. Hornig thought the president would need more information about the "expected military effect in quantitative terms." Yes, the basin was "strategically situated," and "optimistically the operation could have major effects," but what would the practical effects be on the North Vietnamese forces? Hornig did not accept that flooding would not be a major hazard for civilians: it would likely be a "hazard to life, health, and sanitation." Losing the rice crop and good transportation would "likely cause food shortages for the very young, the aged, and the infirm." Hornig concurred that a "leak" was unlikely, but argued that it would be compromised at some point for one of three reasons. Saigon radar was already picking up "unusual cloud formations," but those observations had been waved off by "ascribing them to 'early monsoon conditions.'" The seeded clouds were so distinctive that they could be seen by nonmilitary weather satellites and meteorologists might be intrigued enough to start doing research. (Military weather satellites were far superior to civilian satellites, able to pick up features that showed up as indistinct blurs on other weather satellite images.) Furthermore, Hornig wrote, "The fact that we are going ahead with the Indian program on the basis of apparently flimsy back-up evidence has led to speculation that we 'know something' which has not yet appeared." Rostow, Hornig declared, was underestimating the "degree of revulsion to be expected in the domestic and international meteorological circles at the initiation of 'weather warfare' . . . especially since our highly disaffected general scientific community, at least in the universities, is ready to join the chorus." Rostow should also be considering "potential damage to our world position" by initiating a "new form of warfare." "Not only because it is in the sequence—atomic bomb, riot gases, defoliation, napalm," Hornig stressed, "but because of the picture it may give of a nation flailing out with every tool at its disposal—particularly if it should prove ineffective."[71]

Hornig was dead-on in his assessment of the revulsion in domestic "meteorological circles," which had been very clear about using weather control only for peaceful purposes. Earlier in the decade, the National Academy of Sciences' Committee on Atmospheric Sciences had rejected Edward Teller's idea that it should propose that NATO study weather control for the very reason that it could jeopardize international cooperation in the atmospheric sciences that had grown despite the Cold War. Teller withdrew his proposal after heated opposition from a number of committee members, including numerical weather prediction pioneer Jule Charney. But he was not easily discouraged. Teller later recommended using weather as a weapon in Vietnam after hearing NOTS personnel claim they could muddy up the Ho Chi Minh trail with their cloud-seeding techniques.[72]

Despite Hornig's unease, weather control missions were conducted in Laos, North Vietnam, South Vietnam, and Cambodia, depending on operational needs and intelligence estimates (see fig. 7.2). Due to their sensitive nature, they were reported as weather reconnaissance missions through normal channels. The Joint Chiefs of Staff received weekly updates, and periodic updates went to the Secretary of Defense.[73]

As of June 9, Johnson had not yet decided whether to extend Compatriot into the Song Ca basin. In response to Rostow's query, Johnson put the topic on the agenda for their meeting on June 13.[74] Johnson's White House diary indicates he met with Rostow, Defense Secretary McNamara, Under Secretary of State Katzenbach, former national security advisor McGeorge Bundy, White House press secretary George Christian, and the US ambassador to the Soviet Union, Llewellyn Thompson, during his weekly luncheon. That evening, Johnson spoke with both Rostow and McNamara by phone.[75] What might have been discussed at the luncheon or during the phone calls is unknown. A month later, however, Johnson queried Rostow about a landslide that had oc-

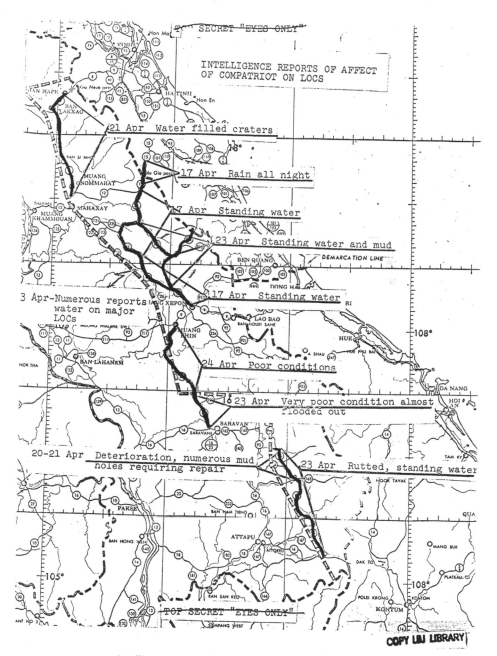

FIGURE 7.2. Intelligence report from Project Compatriot, showing seeding areas and ground conditions, Project Compatriot Weekly Report #6, April 21–27, 1967. National Security File, Subject File Addendum, Box 52/Project Compatriot, Lyndon Baines Johnson Presidential Library.

curred in the A Shau valley, trapping North Vietnamese forces that US forces bombed with B-52s. Rostow reported that the president's instincts about the slide had been correct: seeding in the area June 16–22 had likely contributed to it.[76] Although weekly Compatriot reports through July 13, 1967, reported no seeding in the Song Ca area, maps of the seeding areas provided by the Department of Defense during the Weather Modification hearings before the Senate Subcommittee on Oceans and International Environment of the Committee on Foreign Relations in March 1974 show that the Song Ca basin was added to the seeding area on July 11, 1967 (see fig. 7.3). The next year, the North Vietnam section of the seeding area was deleted, but otherwise the Song Ca seeding area requested by the Joint Chiefs of Staff in May 1967 remained in place until 1972, when seeding moved south into Cambodia.[77] That the Song Ca area was added one day after Rostow confirmed Johnson's hunch that the A Shau valley landslide was connected to Compatriot is not exactly a "smoking gun," but it

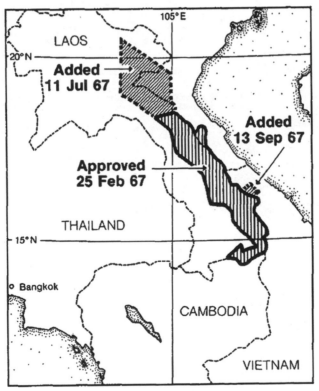

FIGURE 7.3. Project Compatriot seeding areas, 1967. From E. M. Frisby, "Weather Modification in Southeast Asia, 1966–1972," *Journal of Weather Modification* 14 (1982): 1–6.

suggests that being able to tie seeding directly to a successful military outcome was a deciding factor for Johnson.

Compatriot missions continued from 1967 through 1972 during the summer monsoon season (March through November). The Air Force and Navy flew 2,602 sorties during which they expended 47,409 pyrotechnic units. Were they successful? Testimony given at the March 1974 Senate hearings was mixed. According to the statement of Dennis J. Doolin, deputy assistant secretary of defense (East Asia and Pacific affairs), the results were "limited and unverifiable." Project Compatriot had been undertaken because it was relatively inexpensive (the planes were there already) and the increased rain seemed to slow down traffic on primitive roads carrying supplies into South Vietnam. Doolin pointed out that cloud seeding worked only when the clouds in the area would have rained anyway—the situation in Southeast Asia—and that interdiction of this sort made sense only in an area where communications were primitive. Since both of these factors held, DoD took advantage of it.[78] But Lieutenant Colonel Ed Soyster provided more specific information. Despite a lack of ground measurements, the Defense Intelligence Agency had estimated a 30 percent increase. Soyster continued, "Sensor recordings and other information following seeding indicated enemy difficulties from heavy rainfall." Defense concluded that seeding had slowed down traffic on the Ho Chi Minh trail.[79]

At the same hearing, Saint-Amand took great pains to describe how good the NOTS-developed cloud-seeding techniques were and how they had been used to relieve the Bihar drought (without saying it was a classified project), mitigate a drought in the Philippines (a project known as GROMET II), produce water in the Azores, and help out the island of Okinawa (then under US military control). When Senator Claiborne Pell asked if Saint-Amand had been involved with weather modification in Southeast Asia—the subject of the hearing—Saint-Amand replied: "I am appearing here as a private citizen. I am not authorized to express any opinions whatsoever one way or the other on the subject, gentlemen. I must decline to answer."[80] But Pell didn't ask Saint-Amand his opinion—Pell asked if he had shared NOTS-developed technologies with those doing seeding in Laos and Vietnam. And he continued to claim that everything NOTS shared was the same as that used in Project Stormfury. This was not true. The devices were not the same, they were classified, and they continued to be modified throughout the Vietnam War to meet new military needs.

Before closing this chapter, I want to address the issue of the secrecy surrounding military involvement with weather control. Very few people—uniformed or civilian—knew much about GROMET, Popeye, or Compatriot. That was the way the Departments of Defense and State wanted it. During his 1974 hearings, Senator Pell asked Dennis Doolin—who passed the question to Air Force major general Ray Furlong—for the "reasoning behind it being so highly classified." Furlong gave a classic non-answer: perhaps there was political concern. When they started interdiction efforts in Laos, they did not discuss them with anyone except the Government of Laos. But, Pell continued, "the classification was considerably higher. . . . This was the only program about which the DOD did not feel able to respond to questions in either a public or private session." Since Pell thought the weather modification scheme had actually produced nothing of importance, he was reminded of the old maxim, "An elephant labored, and a mouse came forth." So, he pressed, why the great secrecy? General Furlong responded that Pell's elephant maxim reflected "in large measure [the] current thinking of the classification." Doolin jumped in: "We are actively pursuing this in terms of declassification of the information."[81] Were they? The documents related to Compatriot in the LBJ Library were not declassified until 1994—twenty years after the hearings. While Doolin may have wanted them declassified, it is hard to know what he did and did not get declassified. But Pell and his committee did not find out who was involved in approving the cloud-seeding areas in Vietnam or who had known about the project while it was being considered and while it was underway . . . or if they did, they were sworn to secrecy as well. And when a few minutes later Pell asked the DoD team if any other classified weather modification projects had been undertaken by the government in the previous ten years or if this were the only one, these were the responses:

GENERAL FURLONG: To the best of our knowledge.
LT COL SOYSTER: It is the only one, to my knowledge.
MR. DOOLIN: The only one.[82]

Were they lying, uninformed, misinformed, or some combination thereof?

And what of the CIA's role in classified weather modification? George Brown claimed that the CIA was involved at some level in the GROMET operation, but if it was, the CIA has not declassified those records and posted them to CIA CREST—its database of declassified documents—as of this writing. Available documents show that the CIA had been tracking Soviet and Chinese weather modification since the 1950s, and that it was interested in potential military applications. In June 1972, a staff member from the Senate Foreign Relations Committee asked the CIA, on behalf of Senator Pell, to provide a "brief survey of weather modification in Indochina." An unidentified CIA officer reported in the June 27, 1972, Journal–Office of Legislative Council that he had told the staffer he knew nothing more about weather modification than was contained in Senator Pell's press release. He further suggested that the staffer talk to the State Department.[83]

But it is unlikely that the CIA was in the dark about military weather control. In January 1973, Rear Admiral L. W. "Bill" Moffit, of the Office of the Chief of Naval Operations, asked the commander of Naval Intelligence Command for U-2 support to provide reconnaissance for weather modification experiments that would be undertaken by the Naval Weapons Center, China Lake (formerly NOTS) over the Pacific Ocean just west of Santa Barbara, California, from January through April. The request was problematic because the previous year, the desired natural conditions under which the experiments would take place had occurred only sixteen times, and Go–No Go decisions were made only twenty-four hours in advance. Any assistance would be appreciated.[84] Less than two weeks later, the Naval Intelligence Command forwarded the original request to the chief of Reconnaissance Group Information Requirements Staff, Office of the Deputy Director for Intelligence, at the CIA, pointing out that "the purpose of the experiments is to develop effective cloud dispersal techniques which, if successful, could have operational applications." The U-2s would be ideal. If the CIA could help out, it should contact the appropriate Navy office.[85] A little less than a month later—apparently after some internal routing of the request within the CIA and the visit of "representatives from Project Headquarters to the Naval Weapons Center"—the chief of the Collection Branch Reconnaissance Group was advised that it would be supporting the Navy on this project, which was given a code name that was redacted from the document.[86] Here we have the Naval Weapons Center conducting experiments off the coast of Santa Barbara. That in itself was not strange, since the "Santa Barbara experiments," as they were dubbed, had been underway for a number of years. But when Admiral Moffit made the initial request for U-2 support, the first paragraph contained a brief description of the experiment, indicating the general vicinity (off the coast of Santa Barbara), the period (January through April), and how far in advance they could predict satisfactory conditions (six hours with 80 percent reliability). That paragraph had a classification marking, and the CIA redacted it and every other classification marking in this secret document. Why? The correspondence that forwarded the original memo to the CIA was similarly redacted. One might expect that original paragraph was unclassified (U), and the paragraph asking for U-2 support was secret (S)—which would still make the entire document secret—but it is impossible to discern. And if the first paragraph were indeed classified secret, then it makes the Naval Weapons Center's work off the Santa Barbara coast even more interesting.

A third example from declassified CIA documents provides additional clues to CIA involvement. In January 1976, Senator Pell was again holding hearings on weather modification. A military officer from the Department of Defense Research and Engineering Office—which had been involved in GROMET—was sent to testify before Pell's Senate Foreign Relations Arms Control Subcommittee. He had contacted the CIA because he expected "to be asked if the CIA" were involved. Several CIA personnel went to work on the request and "worked out some language" for the officer's use.[87] That "language" was probably provided over the phone to avoid leaving a paper trail, and as it turned out, Pell did not ask the DoD team about CIA involvement during the January 21, 1976, hearing.[88] But in July 1976, Pell wrote directly to then CIA director George H. W. Bush and asked point blank: did the CIA "conduct, finance, or encourage other groups . . . to seed the wind currents and clouds off the coast of Cuba (or any other nation) to cause them to drop rain before they passed over that country" between 1968 and 1972? Or did it seed clouds "to cause severe storms or flooding that would impede the harvest activities within that nation"?[89] The letter enclosed an Associated Press article that had appeared in the *Providence Journal* on June 29, which included the original allegation, a denial from the Department of Defense, and a sentence indicating the CIA could not be reached for comment.[90]

But CIA director Bush was already ahead of the game. On July 7, a letter to the *New York Times* written by a University of Maine history professor discussed how a "former Defense Department researcher revealed that the United States has been waging a secret weather war against Cuba," and tied that weather war to the CIA. At the bottom of the article is a handwritten note, "Ben—Any dope on this? GB 7–7"[91] The reply to Bush's query came nine days later: after checking files and "institutional memory . . . we have never been involved in any operations directed to weather modification over Cuba. There is an old feasibility study done by the DDS&T [deputy director for science and technology] re the possibilities, but the project was abandoned."[92] The letter drafted by CIA staffer G. A. Carver, Jr., for Bush to send to Pell was a short two sentences denying CIA involvement.[93] However, Carver provided Bush with a second paragraph to add to the original draft:

You will recall that in connection with our correspondence earlier this year (your 23 January letter to Bill Colby and my 19 February reply) my colleagues briefed two members of your personal and committee staff on certain Southeast Asian activities, none of which were directed against harvests or—obviously—against Cuba.

Carver was inclined to send just the one paragraph answer and

convey the flavor of the above orally through a telephone call or in your next face-to-face meeting. That would keep our statements in the public written record accurate but avoid whetting curiosity. On the other hand, you may not care to sail quite this close to the wind so far as the written record is concerned and, hence, may want to incorporate something like the suggested second paragraph in the actual written response.[94]

In his August 16 response, Bush—who would become Ronald Reagan's vice president four years later, and US president eight years hence—elected to include the second paragraph in his letter and finished by writing, "I would be happy to discuss this further with you personally if you wish."[95]

By the mid-1970s, Senator Pell's public hearings had brought out what the Department of Defense was willing to report about weather modification. Members of the Pentagon staff were careful to reassure the senators that DoD was not working on any classified research related to environmental modification of any sort, be that seeding rain clouds, clearing fog, steering typhoons, or dropping emulsifiers on roadways. The CIA responses, based on the "sailing too close to the wind" comment, suggest the possibility that the CIA—at least in its unclassified public comments—was shading the truth. What it was willing to report behind closed doors will remain a mystery until the records are declassified, if ever. But in any case, the outrage over weather modification had put a damper on the US government's attempts to use weather control as a military tool, and perhaps as a diplomatic one as well.

Domestic programs were still in effect, but they were starting to run into headwinds as well. As we will see in the conclusion, the 1980s would not be kind to government-sponsored weather control. But the concept of weather control, like a vampire, is difficult to kill.

The weather control slide that began in the 1970s was almost as steep as the weather control climb that started in the late 1940s. On the military side, the "outing" of the secret weather modification project in Laos and Vietnam by *Washington Post* columnist Jack Anderson in 1971, its inclusion in the *Pentagon Papers*, and then the full-blown exposé by journalist Seymour Hersh in the *New York Times* in July 1972 were just a warm-up before an outraged Senator Pell started his hearings: the first, on Senate Resolution 281, "Prohibiting Military Weather Modification," in July 1972 and then the 1974 hearings discussed in chapter 7.[1] The intense scrutiny put the Department of Defense on the . . . ah . . . defensive, and the personnel trotted over to Capitol Hill to testify must have drawn the short straws. From that point on, weather control took on an even lower profile in the military. By fiscal year 1978, the military branches zeroed out the weather modification R&D budget.[2] However, those numbers were for unclassified R&D. And while in the late 1990s the US military was "[virtually ignoring] current techniques of weather modification," the idea of using weather control for military operations was not completely dead. Fog suppression, for example, was still ongoing at Fairchild Air Force Base in Spokane, Washington, though only on a limited basis. Base officials were reluctant to use it because the technique left nearby roads, including those outside the base, icy. Base commanders did not want to be legally liable for a traffic accident.

However, in his thesis "Benign Weather Modification," Air Force major Barry B. Coble argued that it was time for the military to take another look at what he called "benign weather modification"—the use of techniques that could cross over into the civilian sector (including fog dispersal and rain enhancement). If it made use of "remotely piloted vehicles" (RPV), that is, drones, which could "loiter" with their seeding payload in clouds at less expense than could piloted aircraft, then there could be a variety of positive outcomes.[3] The military services performed such "benign" missions during Project GROMET in India, and later GROMET II in the Philippines, but spending time, money, and energy on benign weather control in an era of high-tech weaponry also delivered by drones seems high unlikely.

On the domestic front, funding decisions were not playing out well for the various weather control–related programs. As discussed in chapter 6, the Agriculture Department pulled funding for Skyfire from the fiscal year 1976 budget. Stormfury was having its own set of problems, not least of which was having no hurricanes that could be safely seeded in the Atlantic and international opposition to seeding typhoons in the West Pacific, which stopped seeding before it started. Skywater was still in play for the next few years, but it ran into trouble too: the dry years of the 1960s abated, removing the urgency from weather control efforts. Despite a reduction in support, operational and research-based weather control projects did continue in the Continental United States and in Alaska during the 1970s (see fig. P3).

However, relatively good weather, problems finding suitable tropical systems for seeding, and sufficient water in reservoirs would not have been enough to remove weather control from the budget. So what else was going on? An excellent assessment was done in the late 1980s by Stanley A. Changnon, Jr., a highly respected atmospheric scientist with the Climate and Meteorology Section of the Illinois State Water Survey who had long been active in weather modification research, and political scientist W. Henry Lambright, an assessment which sheds some light on the rapid fall of weather modification funding. Although $300 million in federal funding had been spent on weather control research since 1960, by the late 1970s funding started to fall rapidly, and by 1987 (the year of their article) it had fallen by half since 1978.[4] Interestingly, funding dwindled precisely when scientists were making significant advances. By the mid-1970s, it was possible to dissipate cold fogs and enhance snowfall from orographic clouds. Furthermore, technological improvements in weather radar, aircraft, and seed-delivery systems pointed to even more progress in the future. Despite recommendations for increased funding by the scientific community, federal funding for weather modification R&D dropped 24 percent in fiscal year 1974 while the rest of the R&D budget increased 11 percent.[5] Why?

At the time, several possible reasons were bandied about, some of which were the "usual suspect" issues that had cropped up in earlier battles for

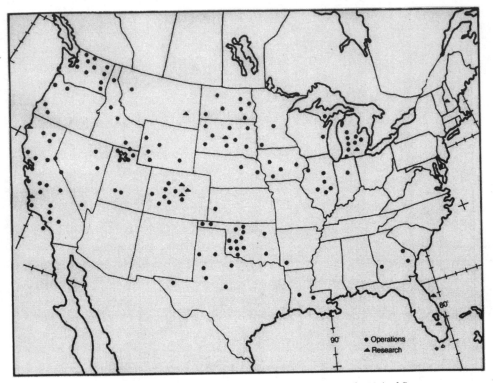

FIGURE P3. Research and operational weather control projects in the United States, 1973–1978. Two additional projects were underway in Alaska. From "The Management of Weather Resources: Report of the Statistical Task Force to the Weather Modification Advisory Board, Vol. 1," Department of Commerce, June 30, 1978.

funding: lousy experimental design, unscientific techniques, poor management, and, perhaps the most damaging charge of all, claims of success that were not backed up by acceptable evidence. But Changnon and Lambert posit a different explanation: a series of federal policy decisions, starting with NSF's removal as the lead agency for weather modification research in 1968, had led to an uncoordinated effort on weather modification research that ultimately doomed the field. Further, they argue that when the NSF shifted management of weather modification activities from its Meteorology Program—where it logically should have remained—to its new Research Applied to National Needs (RANN) Program, the focus shifted from the basic atmospheric research that scientists still needed to undertake in support of weather control to applied research, succumbing to the temptation to just get on with enhancing rainfall or clearing fog or whatever needed to be done to meet "national needs." This focus shift coming at the same time as the cancellation of the National Hail Research Experiment, the weather control research budget at NSF dropped from $5.4 million in fiscal year 1976 to $2.0 million in fiscal year 1978. Although the program later turned back to basic research, by fiscal year 1982, the annual research budget for weather control was only $1.0 million—mist in the rain gauge for a technology-heavy experimental field.[6]

Even though NSF's drawdown on weather control funding was a huge loss, other agencies might have taken up the slack. They didn't. Agriculture zeroed out Project Skyfire's budget by fiscal year 1977, but considering that agriculture could have been *the* major beneficiary of a successful national weather control effort, the department could have shown more interest all along. It appears that Agriculture had decided to defer to other agencies—for example, the Bureau of Reclamation (BuRec)—to make that investment. High crop yields during this period would not have encouraged Agriculture to divert funds to a high-technology method of providing water for crops when the crops were doing just fine.[7]

What about Interior, BuRec, and Project Skywater? Western senators, who were always interested in getting more water to their constituents, had been early supporters of Skywater and had appropriated increasing sums of money in the mid-1960s. But while BuRec did conduct, or contract for, basic atmospheric research, most of the funding went to applied research aimed at putting water into watersheds that fed the existing reservoir/irrigation system. None of their basic research funding addressed the possibility of mitigating severe weather—hurricanes, tornadoes, or hailstorms. That meant that Interior/BuRec would not become the lead agency for federal weather modification efforts, despite support on Capitol Hill. Lead agency or not, BuRec ended up getting the most federal support for weather control, skewing federal policy toward weather control as a water resources management tool.[8]

The National Oceanic and Atmospheric Administration (NOAA), which in 1970 became the umbrella organization of the National Weather Service (formerly the USWB), would have been the logical lead agency for the weather control effort since the late 1940s, but USWB chief Francis Reichelderfer did not support it. When he retired in 1963, Robert M. White, who was more favorably disposed to weather control research, took his place. (White later led the Environmental Science Services Administration and then NOAA.) With White at the helm, the USWB re-entered the weather control research arena with Project Stormfury. Additional weather control projects included the Florida Area Cumulus Experiment (FACE), which started in 1970 and continued throughout the decade, efforts to "redistribute lake-induced heavy snowfalls on Lake Erie," and lightning suppression in Arizona. For reasons that are not clear, NOAA's leaders did not seek to take the national lead on weather control.[9]

In any event, with the military services knocked out of the running after the weather weapon fiasco of the early 1970s, and the other agencies either uninterested or unable to take on a leadership role, no one stepped forward to press the case for sustained federal funding for weather control. What had been considered to be in the nation's interest—economically, socially, and militarily—had been pushed aside. The possibilities for its resurrection were anything but clear.

Weather Control and the American State

Since there are multiple ways to see this story, we need to consider multiple perspectives. Therefore, the conclusion considers episodes in the history of weather control and its relationship to the American state, each of which contribute to the larger picture.

The American state, bureaucratically strong or not, has always meant "officials doing things"—whether we have in mind an administrative, industrial, welfare, national security, warfare, proministrative, associational, scientific, straight, or any other kind of state.[1] And those officials, like state officials everywhere, exerted control over the state's territory and residents. But starting in the late nineteenth century, state officials attempted to exert more control over natural resources by manipulating them on scales large and small. Forests and grazing lands; all manner of flora and fauna, terrestrial, aquatic, and atmospheric; the soil; minerals; and especially water for domestic consumption, hydropower, irrigation, industrial processes, transportation, and recreation—all came under state scrutiny in aid of the military, economics, and public health. Officials' efforts to control or merely subdue nature were aided by scientific knowledge, at first imported from Europe and eventually developed in domestic research university, government, and industrial laboratories. But most of it boiled down to tinkering with nature, since they were manipulating processes underpinned by limited scientific understanding. These activities eventually became central to bureaucratic decision making. Engineers and scientific experts were called upon to advise the state, carry out appropriate research and applications with state funding, and professionalize the state's operations.[2]

But even in a continent-sized country like the United States, controlling these types of natural resources through the first half of the twentieth century was "local" because actions and consequences remained within state territory. Managing forests throughout the Pacific Northwest is a substantial undertaking, but that control does not extend to Canadian trees. Tailings resulting from mining in the Intermountain West may adversely affect people living downstream from the mines, but probably not people living in South America. Similarly, ridding the land of predator species in an attempt to increase the number of "useful" animals does not affect people or ecosystems in Asia. All of these instances represent local control, even if they cover geographic regions that are larger than many other nation-states.

But controlling the behavior of air and its moisture (which recognize no political boundaries) and water (which may flow out of one nation-state and into another) raises the stakes significantly if the consequences, unintentional or not, "bite back."[3] A state that attempts to *deliberately* control these two most basic natural resources in order to militarily or diplomatically coerce other nations to do its bidding will quickly run afoul of the international community of nations (as in fact did happen, as we have seen, when the United States weaponized weather in Southeast Asia). And what if a powerful state—for it will only be within the purview of a powerful state—decides to take a global environmental matter into its own hands and try to engineer its way out of it? What then? Such an attempt could trigger a major global power struggle in the twenty-first century.[4]

We now know that in the United States, scientific—loosely defined— weather control was always a state tool. From the first explosives-laden experiments outside Midland, Texas, in 1891, to the Roaring Twenties–era electrified sand experiments to clear clouds and fog for army aviators, to General Electric's post–World War II dry ice and silver iodide cloud-busting and rain-inducing experiments, and on to the deep-in-the-Cold-War applied weather control projects undertaken by the US Forest Service, Bureau of Reclamation, Weather Bureau, and the Department of Defense, American "officials taking action" funded the research, developed and executed the applications, and then decided when the weather control tool was no longer worth using.

All this matters in a new way in the twenty-first century as the effects of climate change are becoming increasingly apparent: glaciers are retreating, sea level is rising, permafrost is melting, plants and animals are either on the move or failing to thrive, fire seasons are longer and more intense, weather extremes may be occurring with more regularity, droughts may be particularly protracted, fresh water may become in increasingly short supply.[5] The American state has picked up and discarded the weather control tool more than once. Will the need for water lead it to pick up this tool again?

The American state—those "officials taking action"—can often be sensed in narratives of US history. It often lurks just at the edges in histories of science, technology, and the environment: a sometime patron, builder, influencer, user, or regulator. So, for the moment, let us imagine the "state" as a shadowy male figure. What if the developments narrated in this history depend on the state— this shadowy man—every step of the way? Do we bring him (and a signed check) in through the front door, then shove him out through the back door so the rest of the actors—scientists, engineers, inventors, business leaders, or others—can remain disinterested and untainted by federal largesse? Or do we pull him in, take his check, and then let him sit in the withdrawing room while everyone else drinks sherry in the main parlor? Or, instead, do we invite him in, take his money, let him "meet the parents," acknowledge that he is calling the shots, and then continue checking in with him to make sure he hasn't trotted off with a more attractive partner whose heartfelt desires are more in sync with his own? If we are going to "bring the state back in" while telling a story about state efforts to control nature in the way that Theda Skocpol encourages, then the last course is the only one that makes sense. We not only need to follow the money; we need to follow the state's motivation for providing those funds and uncover the hoped-for return on the investment.[6]

In the late nineteenth and early twentieth centuries (as discussed in chapter 1), the American state's motivation for using weather control was extremely weak, particularly compared to the sophisticated Cold War uses that came to the fore after World War II. The initial attempt, prompted by a lone senator who wanted to increase the value of his Texas landholdings by having rain fall on his property, was relatively low cost and of limited scientific value. However, it fit in nicely with the invention-crazy milieu of the time when nothing was impossible, not to mention the state's desire to make something of that huge expanse of land between the Mississippi River and Pacific Ocean, from the Canadian border south to the Mexican border. That effort died quickly. The Agriculture secretary—seeking to advance his own agenda for turning his science-heavy department into a bureaucratic powerhouse—disowned the entire project and returned the money to the US Treasury. Nevertheless, the initial project would not have been undertaken without the efforts of the state, and the Progressive Era push to increase efficiency and rely on rational, scientific modes of "taking action" paved the way for additional congressional appropriations for scientific investigations that would allow the state to control its natural resources.

The state also inspired the rise of aviation as a military tool, and World War I's impact set the stage for the next attempt: using electrified sand to clear fog and clouds for fragile Army Air Corps aircraft whose pilots, in the days before instrumented flight, relied on visual flight rules. They had to be able to see the landing strip before trying to bring their aircraft down for a landing, or, more accurately, a controlled crash. Clearing clouds or fog with electrified sand allowed them to do so. Alas, this early twentieth-century state effort also came to nothing. The idea was attractive—the execution was not. But again, it was totally dependent on state funding to develop the instruments for electrifying the sand and military aircraft to carry out the experiments.

The development of a theoretical basis for precipitation processes in the 1930s helped pave the way for the first genuinely scientific efforts to control the weather, which grew out of General Electric's research on cloaking smokes and aircraft icing during World War II. The state's importance becomes increasingly evident here. The laboratory director, Nobel Prize–winning chemist Irving Langmuir, had long gotten funding from the US military. To secure funding for weather control, all he needed to do was make a few phone calls or send a letter to his defense contacts, and the deal was made. Not only did he get funding: he got access to military airplanes, airfields, and other instrumentation required to carry out his experiments because military leaders could see many ways to use it as a military tool in the early Cold War. It could not be traced back to the user, it wasn't radioactive, it could be used tactically and strategically, and it was relative inexpensive. What was not to like?

Within a few months, interest spread to scientific experts, bureaucrats, and congressmen; the state hold on weather control tightened. In weather control, congressmen, particularly New Mexico senator Clinton P. Anderson, were pursuing goals that did not seem to reflect much more than a postwar desire to simultaneously control nature and bolster national security, both domestically and militarily. He wanted not only to advance weather control as a state tool by building a bureaucratic structure around it (Weather Control Commission), but to maintain it as a state tool by keeping commercial interests under federal control. But did the state have the capacity to implement such a goal? Certainly the money was available—and so was the bureaucratic infrastructure, which had not just expanded during the war, but had solidified into place as the Cold War deepened.[7] Pockets of opposition pushed back, including the military—which wanted to keep weather control out of the hands of another Atomic Energy Commission–type bureaucracy so that it could maintain control—and commercial seeders, who wanted to be able to pursue their business interests without state interference. Ultimately, the military kept control and the commercial seeders were reined in by individual states. That is, they experienced sub-state control.

Scholars have argued that the American state does not have the typical underpinnings that support strong state power, such as the French state's prestigious career civil service, and therefore its power is "fragmented" and "dispersed."[8] With weather control, that is certainly the case. Yes, the military had a fairly tight hold on weather control research, but the US Weather Bureau in the 1950s and the Bureau of Reclamation in the 1960s also administered weather control *funding*. Did that really make a difference to recipients of federal grants to conduct related research? When funding is coming from federal coffers, it may seem to researchers that their research agenda is being steered by the state. Yet, when one considers the development of scientific expertise and the execution of scientific research in the United States, the fragmentation may not make as much difference as it does in other situations. The money comes from different agencies, but it is all under state control.

This also raises several intriguing questions. How, for example, did the state try to seek optimal solutions to problems? One should also not dismiss the possibility of state actions addressing problems or finding solutions beyond the "reach of societal actors."[9] Indeed, if one is trying to figure out how to control the weather, that is not going to be done by people in society. Researchers may figure out the mechanics, but they are not going to address national issues, much less international implications. One may also argue that no state activity can be "disinterested" because someone or some group will benefit and someone or group will not.[10] Indeed, how does the state find solutions when there are no disinterested actors? In the case of weather control, it was not triggered by anyone who was disinterested—nor was it opposed by anyone who was disinterested. Everyone concerned had a vested interest, whether they were within or outside the state apparatus. Langmuir and his group supported weather control because it was good for General Electric even as they claimed it was good for the nation. Western farmers and ranchers who supported it were looking for cheap water, so they weren't disinterested. Congressmen who simultaneously wanted more research and more applied weather control along with more control over weather control were not disinterested—they very much wanted their states to be the beneficiaries of weather control successes. The military and the Bureau of Reclamation both wanted weather control. And the weakest actor in this group—the Weather Bureau—wanted it to just disappear. But disinterested parties? Not a one.

We might also ask: in practical terms, what were some of the primary management problems facing weather control in the first decades of the Cold War?

Concentrated efforts shifted from basic research to practical applications too soon, and one of those practical applications had been the weather weapon, the exposure of which during the Vietnam conflict left the United States redfaced. Furthermore, commercial seeding firms were continuing to provide weather control services—thus making it appear that the time for basic research was past, and it was time to move on to applications. Many scientists, however, did not agree. This led to public disagreements *among* meteorologists over the efficacy of weather control projects, *between* meteorologists and engineers over the efficacy of weather control technology, and *between* users of weather control services (who wanted to see seeding projects expand) and scientists (who were not sure that was a good idea). As uncertainties increased, willingness to extend or expand funding decreased. Also at issue was the limited number of field projects, which tended to be underfunded and poorly designed, leading to disputes over the statistical results. Consistently inconclusive results led many scientists to conclude that "poor science" was involved—an almost certain death knell for a scientific undertaking. And lastly, there were questions concerning whether scientific gains that had come from the $300 million expended on weather control R&D had been worth it. Could the money have been better spent on research that would have improved forecasts or for more sophisticated atmospheric models? That is impossible to determine—there is no opportunity for a "do over" with research monies. But once the questions started being asked, the lack of answers led scientists to pull away and encourage the flow of research dollars into other areas of atmospheric sciences.[11]

"The past," as William Faulkner notes, "is never dead. It's not even past."[12] Or as Yogi Berra put it, "It's déjà vu all over again." Scientists may have been retreating from weather control research as the money dried up by the end of the Cold War, but does that mean that federal legislators had permanently turned their backs on weather control? Not at all.[13] After the 1980s, weather modification and control disappeared from the congressional hearings lineup. But in 2005, during the George W. Bush administration, hearings were held on Senator Kay Bailey Hutchison's (R-Texas) Weather Modification Research and Technology Transfer Authorization Act (S. 217), which would have established the Weather Modification Operations and Research Board under the Department of Commerce. Its purpose was to "develop and implement a comprehensive and coordinated national weather modification policy and a national Federal and State program of weather research and development."[14] And why did Senator Hutchison think this was a good idea? According to

Hutchison, it was critical to "assess and evaluate the efficacy of weather modification research to the extent that lives are saved and property damage is limited," especially considering the impacts of hurricanes Katrina and Rita, and violent tornadic storms that had hit the Midwest during 2005.[15] And she does hail from a state that has seen its share of droughts and does have its own Weather Modification Advisory Committee.[16] Scientists did testify in support of the bill—which was favorably reported out of committee, but with no attached funding—arguing that considerable technological progress had been made since the 1980s and more money needed to be made available for research to exploit these new techniques. As amended, the proposed board would become a subcommittee under the Office of Science and Technology Policy to "coordinate a national weather modification research program."[17] In a pattern reminiscent of bills introduced each session during the 1950s—it is unclear if Hutchison was aware of this history—this bill never made it to the Senate floor for a vote.

Two years later, Senator Hutchison tried again with S. 1807, "Weather Mitigation Research and Development Policy Authorization Act of 2007," which would have established the Weather Mitigation Advisory and Research Board. The purpose of the act was to "develop and implement a comprehensive and coordinated national weather mitigation policy and a national cooperative Federal and State program of weather mitigation research and development"— essentially the same as her earlier bill. However, instead of being under the Department of Commerce, the board would be under the supervision of the National Science Foundation. Interestingly, "mitigation" was defined as being "modification"—not an equivalence that scientists are likely to make.[18] Once again, the bill died, was reintroduced (2009), successfully passed committee hearings, but failed to go to the Senate floor. The lack of a major weather disaster coupled with a tight budget in the Great Recession may have helped to doom Hutchison's bill. Or, perhaps, legislators are no longer so keen to spend money attempting to control the uncontrollable.

This notable lack of success in passing legislation and securing funding for weather control did not, however, stop the Department of Homeland Security (DHS) from asking the National Oceanic and Atmospheric Administration (NOAA) in 2009 to help DHS develop its own program to modify hurricanes. But NOAA demurred, responding that its Office of Atmospheric Research recognized "that weather modification, in general, is occurring through the funding of private enterprises," but that NOAA "did not support research" that would entail "efforts to modify hurricanes."[19] So NOAA has stayed away from weather control efforts for decades. Apparently, DHS did not do its homework first.

While no other weather modification–related bills (other than companion House bills) have been introduced at the time of this book's publication, hearings have been held on geoengineering, also known as climate engineering, "to address global climate change problem, including carbon dioxide removal (CDR) and solar radiation management (SRM) technologies."[20] The first hearings took place in November 2009 and continued periodically until March 2010. The Congressional Research Service and the Government Accountability Office have also provided reports related to geoengineering, but they generally have addressed carbon dioxide removal and solar radiation management technologies.[21] Geoengineering, as defined in these publications, is only distantly related to the weather control efforts of the mid-twentieth century. Since most of those were not "large-scale" projects, they would not qualify as geoengineering, which takes human tinkering with the environment to a completely different and sobering level, and involves geopolitical considerations in a way that weather control does not.[22]

Does this mean that weather control is on the skids? On the contrary. It may surprise readers to know that a number of state governments continue to be involved in weather control efforts, including hail mitigation and cloud seeding for snowpack enhancement. Examples of current state-supported weather control projects include North Dakota's hail suppression project, snowpack enhancement projects in Utah, Wyoming, and Colorado, and Colorado's use of "sound cannons" to break up hail-producing storms. North Dakota's program—operational since 1974—seeks to protect property and crops. Sponsored by local county governments with state government sharing the costs, the program is directed by state meteorologists and carried out by Weather Modification, Inc., in Fargo, North Dakota.[23] Utah's snowpack enhancement project is under the direction of the Utah Division of Water Resources and has been ongoing since 1973. In 2014, silver iodide cloud seeding was underway in four areas of the state at a total cost of approximately $500,000. The State of Utah covers half of the cost, with the remainder funded by various water conservancy districts. Based on 2012 data (the latest posted), cloud seeding increased runoff by 181,700 acre-feet of water at a cost of $2.27 per acre-foot, and percentage increases ranged from 3 to 15 percent, depending on the seeding area.[24] From the data, it is impossible to determine whether the increase was statistically significant or not. Similarly, Wyoming's legislature approved a five-year Wyoming Weather Modification Pilot Program that got underway during the 2006–2007 winter season. The $8.8 million program differs from others because its results are being evaluated by atmospheric scientists working at the National Center for Atmospheric Research in Boulder, Colorado, and the Desert Research Institute in Nevada (Reno and Las Vegas). Data collection stopped in April 2014, and a series of related reports have already been released.[25] In Colorado, the state's Water Conservation Board issues grants for silver iodide seeding for snowpack enhancement in watersheds feeding the main rivers whose headwaters are in the state. These mostly involve cloud seeding with silver iodide, but they also permit "sound cannons" for breaking up hailstorms.[26] Other states—particularly in arid and semi-arid parts of the country—have similar programs, and a quick look at the annual edition of the *Journal of Weather Modification* provides additional evidence of significant weather control work underway in the United States. The Weather Modification Association currently lists about twenty certified weather modification operators and another nine certified managers who provide services to governments and business interests.[27] Despite NOAA's long-standing qualms about being involved in weather control, the private sector has no such concerns, continuing both research and applied operations.

And lastly, one cannot address the topic of current-day weather control ambitions without a nod to the conspiracy theories that continue to swirl around it. As Area 51 in Roswell, New Mexico, is to alien landings, military-funded efforts such as the now-closed High Frequency Active Auroral Research Program (HAARP) in remote Alaska—which was part of an ionospheric investigation—are to weather control. While a Google search of "weather control" yields approximately 725,000 hits, and "weather modification" yields approximately 760,000 hits, searching for "weather control" and "conspiracy" yields approximately 890,000 hits. Mix together a little bit of secrecy, an almost complete lack of scientific understanding of atmospheric processes, and a conviction that a nefarious hand must be behind every "weird weather" event, and weather control becomes a prime target for conspiracy theorists. Add in the wide reach of the Internet, and the stories start to propagate faster than well-fed fruit flies. Perhaps the weather control story contains so many you-can't-make-this-stuff-up moments that the temptation to make it an even wilder story is too enticing. The conspiracy stories are entertaining . . . just not terribly likely.[28]

So we must conclude this story. Weather control—when considered as a small-scale effort—first gained significant promise after World War II, aided by the rise of the atmospheric sciences in the postwar years and the availability of funding as the Cold War ebbed and flowed until the late twentieth century. But the hype that surrounded it—hype created and sustained by elite researchers such as Irving Langmuir, John von Neumann, and Edward Teller—and worries over Soviet capabilities to exploit the weather to spread Communism around the world made weather control an irresistible draw for military leaders looking for an undetectable offensive and defensive weapon. Thus weather control efforts fit in with the postwar, Cold War–era hubris that people could gain the power to dominate nature. This was, after all, the same era when electricity from nuclear power would supposedly be too cheap to meter, nuclear weapons could be used to create harbors in the Arctic (radiation be damned), and chemicals could control all manner of pests without harming the environment or causing a silent spring—none of which panned out in the absence of scientific and technological humility.[29] Simultaneously, weather control seemed to offer a relatively simple, low-cost solution to the drought conditions that had spread throughout the United States in the 1950s and continued into the 1960s. To congressmen seeking to bring water to their constituents' parched land—and who also wanted to strengthen America's national security posture, particularly in the face of the Sputnik launch—weather control seemed like an amazing possibility that could come to fruition in a few short years if enough money were thrown at it. Cloud seeding for snowpack enhancement in the West's mountains did become a reality. Yet the idea touted by Irving Langmuir that a single pellet of dry ice carefully placed into a cloud in New Mexico could lead to gentle rainfall all the way to the Eastern Seaboard did not. Neither did hurricane-busting efforts, nor ideas that those folks living in the upper Midwest who were suffering from the winter blues could be given periodic shots of sunshine courtesy of designer weather. However, the "weather weapon" did indeed happen—although just how effective it was is highly debatable. It likely washed out enemy positions and muddied up the Ho Chi Minh trail in the late 1960s and early 1970s, but it certainly did not stop the Viet Cong from infiltrating South Vietnam. Its discovery by a persistent, probing American press, combined with its exposure in the *Pentagon Papers*, stirred up a huge controversy over the weaponization of the weather and the possibility that changes in the weather in one part of our global atmosphere might produce inadvertent changes somewhere else. And although it did not precipitate weather control's demise, the classified weaponization of weather did not help to keep it alive: a kind of Faustian bargain, which John F. Kennedy had cautioned about just before his assassination in a speech to the National Academy of Sciences.[30]

What makes this story even more important? It is well worth emphasizing again: because other attempts to control nature—no matter how big—are essentially *local*. Big levees on the Mississippi, the Central Valley water project in California, large-scale management of forests, controls over fish stocks, and the like: all are local in that they do not, nor will they ever, affect the entire Earth. Yes, they may affect a large hunk of the United States or a substantial swath of the ocean, but they do not affect every piece of land or all of the world's oceans. But once people start attempting to control the atmosphere and the moisture it carries, they are affecting the weather and climate of *everyone* downstream. And if they go really big, and think that by sequestering large amounts of carbon dioxide, or putting up reflective devices to send the sun's rays back into space, or putting iron filings into the ocean to suck up carbon dioxide—among the current geoengineering schemes—that they can mitigate rising global temperatures, then the resulting changes in the atmosphere must be agreed to by the entire global community. It is not up to one scientific and engineering powerhouse of a state to attempt massive changes to an absolutely essential natural resource in hopes of changing the atmosphere back to what it was before human actions began to influence it well before the beginning of the industrial age.[31]

But the weather control discussed in *Make It Rain* was not geoengineering, for it could not change the weather on any kind of a massive scale, Irving Langmuir's claims notwithstanding. Skyfire, Skywater, Stormfury, Popeye, Compatriot, and GROMET were instead all about *water*: water to put out fires, water to fill reservoirs for irrigation, reducing water from hurricanes, water to flush out the enemy in Vietnam and Laos, and water for agriculture in India. The American state used a tool, which it called weather control, but was actually the artificial inducement of precipitation on demand: to bring water where it was needed (irrigation—domestically and internationally), wanted (weapon), or not wanted (overseeding to protect fruit or to disrupt hurricanes), or to dry up clouds (fog dissipation, hail mitigation). Those efforts were essentially *local*.

While meteorologists may hold differing professional views of the efficacy of weather modification techniques, I think most would agree that it is possible to modify clouds in some way with current seeding techniques. Whether it is economically worthwhile to do so depends on the environmental context. But if it were not economically viable, consulting firms that provide weather modification services to private customers would have been out of business, and individual states would have given up on their efforts to use these same techniques to fill reservoirs that provide water for a variety of uses years ago.

Recent attention-grabbing headlines about drought in California and throughout much of the US West are evidence of the impact of regional and local fresh water shortages. While they may eventually be ameliorated by several consecutive rainy seasons, the more likely scenario is that the affected regions will have to fundamentally change the way that water is allocated and used. It is just as likely that residents will demand that "something" be done.[32] And the promise of enough rain or snow to keep crops alive may very well seem doable. As we have seen from earlier attempts to alleviate drought conditions, it is possible to induce precipitation from a cloud, but impossible to create the cloud itself. Will multiyear drought conditions finally yield some straight talk on weather control? Maybe, maybe not. But ultimately, the American state's weather control tool existed to place water where it was needed or wanted. And in the twenty-first century, a thirsty population, a thirsty economy, a thirsty state will need to find supplies of fresh water.

How will that demand for water resources be met? Will the discussions related to state actions focus on science and technology? Or will they also address *values?* Public policy debates about climate change and related ecological changes, coastal inundation, and fresh water shortages aren't really about science—they are about values. Expert advice from scientists is not going to influence cultural attitudes in this arena, but looking back at past efforts and their intended and unintended results may help us to put science policy decisions into a context that helps to move the discussion forward.[33] Rainmaking, rain enhancement, hail mitigation, fog clearing . . . all amount to weather control. Designer weather may be down as a tool of the American state, but it is by no means out. Its history holds lessons for the future if we heed them.

ABBREVIATIONS

ARCHIVE/LIBRARY ABBREVIATIONS

Anderson: Clinton Presba Anderson Papers, Manuscript Division, Library of Congress, MSS 10926

Chaffee: Emory Leon Chaffee Papers, Harvard University Archives, HUG 4274.20

Church: Philemon Edward Church Papers, University of Washington Archives, Manuscript Collection No. 2911

DDE/WHCF: White House Central Files, Dwight David Eisenhower Presidential Library, Abilene, Kansas

Dominy: Floyd Dominy Papers, American Heritage Center, University of Wyoming, Laramie

E/AP/RA: Ecology/Air Programs/Remedial Action, Weather Modification Board Reports, Washington State Archives, Olympia

E/AP/REG: Ecology/Air Programs Division, Regulatory Section—Weather Modification Files 1920–1970, Washington State Archives, Olympia

E/ARM: Ecology Department, Air Resources Management Files, Washington State Archives, Olympia

Houghton: Henry G. Houghton Papers, Massachusetts Institute of Technology Archives, MC 242

NOAACL: National Oceanic and Atmospheric Administration Central Library, Silver Spring, Maryland

Langmuir: Irving Langmuir Papers, Manuscript Division, Library of Congress, MSS 29413

LBJ/Classified: Classified Files, Lyndon Baines Johnson Presidential Archive, Austin, Texas

LBJ/India: National Security Files, Country File: India, Lyndon Baines Johnson Presidential Archive, Austin, Texas

LBJ/NSF: National Security Files, Lyndon Baines Johnson Presidential Archive, Austin, Texas

LBJ/WHCF: White House Central Files, Lyndon Baines Johnson Presidential Archive, Austin, Texas

Magnuson: Warren Grant Magnuson Papers, University of Washington Archive, Manuscript Collection No. 3181

Orville: Howard T. Orville personal papers

RDB: Research and Development Board, Records of the Department of Defense, RG 330, National Archives and Records Administration (NARA) II, College Park, Maryland

Reichelderfer: Francis W. Reichelderfer Papers, Manuscript Division, Library of Congress, MSS 61564

Rosellini: Albert D. Rosellini Papers, Washington State Archives, Olympia

Schaefer: Vincent J. Schaefer Papers, M. E. Grenander Department of Special Collections and Archives, University at Albany, SUNY [accessed before they were processed, hence no box numbers]

SecAg: Records of the Secretary of Agriculture, RG 16, National Archives and Records Administration (NARA) II, College Park, Maryland

State3008D: Records of the Department of State, RG 59, Entry 3008D, File 2, National Archives and Records Administration (NARA) II, College Park, Maryland

State5255: Records of the Department of State, RG 59, Entry 5255, National Archives and Records Administration (NARA) II, College Park, Maryland

USWB: Records of the US Weather Bureau, RG 27, National Archives and Records Administration (NARA) II, College Park, Maryland

Von Neumann: John von Neumann Papers, Manuscript Division, Library of Congress, MSS 44180

Wexler: Harry Wexler Papers, Manuscript Division, Library of Congress MSS 45229

JOURNAL/NEWSPAPER ABBREVIATIONS

BAMS: *Bulletin of the American Meteorological Society*
JWM: *Journal of Weather Modification*
NYT: *New York Times*

NOTES

INTRODUCTION

1. James Rodger Fleming, *Fixing the Sky: The Checkered History of Weather and Climate Control* (New York: Columbia University Press, 2010). Surveying literary accounts as well as real-world attempts to control the weather, Fleming focuses on the innate hubris of such ideas and what he views as the pathological nature of the entire enterprise. I concur in pathological science—as defined by Nobel Prize–winning chemist Irving Langmuir, one of the most prestigious scientists involved in weather control—being an element in Langmuir's attempts. Indeed, Langmuir's weather control is perhaps the quintessential example. But Langmuir was not the only scientist involved in weather control, and the scientific work of others, who will appear later in this book, was not pathological: they recognized the limits of what could be done, transparently operated within those limits, and were making sincere efforts to unlock the secrets of cloud physics. And while the military was more than happy to fund weather control efforts, the domestic use of weather control was just as strong. In all of the cases presented in detail here, the common thread is *the state,* exerting control over the atmosphere just as it had exerted control over the rest of its territory.

2. Scholarly treatments of weather control include Clark C. Spence, *The Rainmakers: American "Pluviculture" to World War II* (Lincoln: University of Nebraska Press, 1980); Jeff Townsend, *Making Rain in America: A History* (Lubbock, TX: ICASALS Publication, 1975), provides an overview through midcentury; William R. Cotton and Roger A. Pielke, *Human Impacts on Weather and Climate* (Cambridge: Cambridge University Press, 1995), take a social science view; Chunglin Kwa, "The Rise and Fall of Weather Modification: Changes in American Attitudes toward Technology, Nature, and Society," in *Changing the Atmosphere: Expert Knowledge and Environmental Governance,* ed. Clark A. Miller and Paul N. Edwards, 135–65 (Cambridge, MA: MIT Press, 2001), provides a short summary of US weather control efforts in the late twentieth century; Jacob Darwin Hamblin, *Arming Mother Nature: The Birth of Catastrophic Environmentalism* (Oxford: Oxford University Press, 2013), addresses military uses of the environment, including weather control.

3. My path to approaching weather control as a state tool was influenced by James C. Scott, *Seeing Like a State* (New Haven, CT: Yale University Press, 1998), and by environmental history literature focused on large control-of-nature projects and the intersection of environmental policy and diplomatic history; see Kurkpatrick Dorsey, *The Dawn of Conservation Diplomacy: U.S.-Canadian Wildlife Protection Treaties in the Progressive Era* (Seattle: University of Washington Press, 1998); Samuel P. Hays, *Beauty, Health, and Permanence: Environmental Politics in the United States, 1955–1985* (New York: Cambridge University Press, 1989); Paul W. Hirt, *A Conspiracy of Optimism: Management of the National Forests since World War II* (Lincoln: University of Nebraska Press, 1994); Ted Steinberg, *Acts of God: The Unnatural History of Natural Disaster in America* (Oxford: Oxford University Press, 2000); Richard White, *The Organic Machine: The Remaking of the Columbia River* (New York: Hill & Wang, 1995); Donald Worster, *Rivers of Empire: Water, Aridity, and the Growth of the American West* (New York: Pantheon Books, 1985); Carmel Finley, *All the Fish in the Sea: Maximum Sustainable Yield and the Failure of Fisheries Management* (Chicago: University of Chicago Press, 2011); Mark Fiege, *Irrigated Eden: The Making of an Agricultural Landscape in the American West* (Seattle: University of Washington Press, 2000). On earlier state attempts to control the weather, see Jeffrey Snyder-Reinke, *Dry Spells: State Rainmaking and Local Governance in Late Imperial China* (Cambridge, MA: Harvard University Asia Center, 2009).

4. Patrick Carroll, *Science, Culture, and Modern State Formation* (Berkeley: University of California Press, 2006), 1. On page 23, he argues that a modern state is an "engineering state."

5. Max Weber, *Economy and Society*, ed. Guenther Roth and Claus Wittich (New York: Bedminster Press, 1968; originally 1922), vol. 2, chap. 9; vol. 3, chaps. 10–13. See also Theda Skocpol, "Bringing the State Back In: Strategies of Analysis in Current Research," in *Bringing the State Back In*, ed. Peter B. Evans, Dietrich Rueschemeyer, and Theda Skocpol, 3–37 (Cambridge: Cambridge University Press, 1985), 7.

6. Alfred C. Stepan, *State and Society* (Princeton, NJ: Princeton University Press, 1978), xii. See also Mark R. Nemec, *Ivory Towers and Nationalistic Minds: Universities, Leadership, and the Development of the American State* (Ann Arbor: University of Michigan Press, 2006), 12.

7. Stephen Skowronek, *Building a New American State: The Expansion of National Administrative Capacities, 1877–1920* (Cambridge: Cambridge University Press, 1982), 3.

8. Ibid., 10.

9. John R. Commons, *The Legal Foundations of Capitalism* (Clark, NJ: The Lawbook Exchange Ltd., 2012; originally New York: Macmillan, 1924), 125; Margot Canaday, *The Straight State: Sexuality and Citizenship in Twentieth-Century America* (Princeton, NJ: Princeton University Press, 2009). Scholarship on the state is voluminous, particularly since it has come under the lens of political scientists, anthropologists, economists, sociologists, and historians, although historical scholarship too often leaves the term undefined. Helpful, in addition to works already mentioned, are Aaron L. Friedberg, *In the Shadow of the Garrison State: America's Anti-Statism and Its Cold War Grand Strategy* (Princeton, NJ: Princeton University Press, 2000), who uses the term "American State" to refer to the executive branch of the federal government; Timothy Mitchell, "The Limits of the State: Beyond Statist Approaches and Their Critics," *American Political Science Review* 85, no. 1 (1991): 77–96, who argues that defining the state "depends on distinguishing it from society"; Richard F. Bensel, "Valor and Valkyries: Why the State Needs Valhalla," *Polity* 40, no. 3 (2008): 386–93, who argues that states have a monop-

oly on violence; Peter B. Evans, Dietrich Rueschemeyer, and Theda Skocpol, eds., *Bringing the State Back In* (Cambridge: Cambridge University Press, 1985), whose edited volume argues that scholars need to reintroduce the concept of the state into their research and writing; J. P. Nettl, "The State as a Conceptual Variable," *World Politics* 20, no. 4 (1968): 559–92; Morton Keller and R. Shep Melnick, eds., *Taking Stock: American Government in the Twentieth Century* (London: Cambridge University Press, 1999); Daniel P. Carpenter, *The Forging of Bureaucratic Autonomy: Reputations, Networks, and Policy Innovation in Executive Agencies, 1862–1928* (Princeton, NJ: Princeton University Press, 2001); Laura S. Jensen, "Government, the State, and Governance," *Polity* 40, no. 3 (2008): 379–85; Brian Balogh, *The Associational State: American Governance in the Twentieth Century* (Philadelphia: University of Pennsylvania Press, 2015); Bartholomew H. Sparrow, "American Political Development, State-Building, and the 'Security State': Reviving a Research Agenda," *Polity* 40, no. 3 (2008): 355–67; Morton Keller, *America's Three Regimes: A New Political History* (Oxford: Oxford University Press, 2007), and "Looking at the State: An American Perspective," *American Historical Review* 106, no. 1 (2001): 114–18; Desmond King and Robert C. Lieberman, "Finding the American State: Transcending the 'Statelessness' Account," *Polity* 40, no. 3 (2008): 368–78; Alan Brinkley, *The End of Reform: New Deal Liberalism in Recession and War* (New York: Vintage, 1995); Jurgen Schmandt and James Everett Katz, "The Scientific State: A Theory with Hypotheses," *Science, Technology, and Human Values* 11, no. 1 (1986): 40–52; James T. Sparrow, *Warfare State: World War II Americans and the Age of Big Business* (Oxford: Oxford University Press, 2011); James Gilbert, *Designing the Industrial State: The Intellectual Pursuit of Collectivism in America, 1880–1940* (Chicago: Quadrangle Books, 1972); and Adam Rome, "What Really Matters in History: Environmental Perspectives in Modern America," *Environmental History* 7, no. 2 (2002): 303–18, who sets up a framework for the "environmental-management state."

10. Skocpol, "Bringing the State Back In," 9–13, Krasner quoted on 13; Carpenter, *Forging of Bureaucratic Autonomy*, 14.

11. Kristine C. Harper, *Weather by the Numbers: The Genesis of Modern Meteorology* (Cambridge, MA: MIT Press, 2008), 52.

12. Skocpol, "Bringing the State Back In," 17.

13. On the early days of the US Weather Bureau, see Gustavus A. Weber, *The Weather Bureau: Its History, Activities and Organization* (New York: Appleton, 1922). On the Weather Bureau through the 1950s, see Donald A. Whitnah, *A History of the United States Weather Bureau* (Urbana: University of Illinois Press, 1961). For the Weather Bureau/National Weather Service's role in numerical weather prediction, including their continuing funding problems, see Harper, *Weather by the Numbers*.

14. Yaron Ezrahi, *The Descent of Icarus: Science and the Transformation of Contemporary Democracy* (Cambridge, MA: Harvard University Press, 1990).

15. For a sampling of literature on science, technology, and state relations, see Carroll, *Science, Culture, and Modern State Formation*; Dominique Pestre, "Science, Political Power, and the State," in *Science in the Twentieth Century*, ed. John Krige and Dominique Pestre (Amsterdam: Harwood Academic Publishers, 1997), 61–76; Mark H. Brown, *Science in Democracy: Expertise, Institutions, and Representation* (Cambridge, MA: MIT Press, 2009); Scott Frickel and Kelly Moore, eds., *The New Political Sociology of Science: Institutions, Networks, and Power* (Madison: University of Wisconsin Press, 2006); Stuart W. Leslie, *The Cold War and Ameri-*

can Science: The Military-Industrial-Academic Complex at MIT and Stanford (New York: Columbia University Press, 1993); David Edgerton, *The Shock of the Old: Technology and Global History since 1900* (Oxford: Oxford University Press, 2007); Jessica Wang, *American Science in the Age of Anxiety: Scientists, Anticommunism, and the Cold War* (Chapel Hill: University of North Carolina Press, 1999); Richard M. Merelman, "Technological Cultures and Liberal Democracy in the United States," *Science, Technology, and Human Values* 25, no. 2 (2000): 167–94; Chandra Mukerji, *A Fragile Power: Scientists and the State* (Princeton, NJ: Princeton University Press, 1989); David J. Tietge, *Flash Effect: Science and the Rhetorical Origins of Cold War America* (Athens: Ohio University Press, 2002); John Krige, *American Hegemony and the Postwar Reconstruction of Science in Europe* (Cambridge, MA: MIT Press, 2006); Lawrence Busch, *The Eclipse of Morality: Science, State and Market* (New York: Aldine de Gruyter, 2000); Naomi Oreskes and John Krige, eds., *Science and Technology in the Global Cold War* (Cambridge, MA: MIT Press, 2014); Paul R. Josephson, *Resources under Regimes* (Cambridge, MA: Harvard University Press, 2004).

16. Sheila Jasanoff, "Ordering Knowledge, Ordering Society," in *States of Knowledge: The Co-production of Science and Social Order*, ed. Sheila Jasanoff (London: Routledge, 2004), 21, 14.

17. Stanley Aronowitz, *Science as Power: Discourse and Ideology in Modern Society* (Houndmills, UK: Macmillan, 1988), 351, argues that science is always subordinate to the state, no matter what form the state takes.

PART ONE

1. Quoted in Donald J. Pisani, *Water and American Government: The Reclamation Bureau, National Water Policy, and the West, 1902–1935* (Berkeley: University of California Press, 2002), xiii.

2. Edward W. Byrn, "The Progress in Invention during the Past Fifty Years," *Scientific American* 75 (July 25, 1896): 82–83.

3. Thomas Parke Hughes, *Changing Attitudes toward American Technology* (New York: Harper & Row, 1975), 5.

4. Ibid., 9.

5. Philip Scranton, "Technology, Science and American Innovation," *Business History* 48, no. 3 (2006): 312.

6. Larry Owens, "Science in the United States," in *Science in the Twentieth Century*, ed. John Krige and Dominique Pestre (Amsterdam: Harwood Academic Publishers, 1997), 812; Mark R. Nemec, *Ivory Towers and Nationalistic Minds: Universities, Leadership, and the Development of the American State* (Ann Arbor: University of Michigan Press, 2006), 2; Jessica Wang, *American Science in an Age of Anxiety: Scientists, Anticommunism, and the Cold War* (Chapel Hill: University of North Carolina Press, 1999), 5.

7. Nemec, *Ivory Towers*, 6–7; Brian Balogh, *The Associational State: American Governance in the Twentieth Century* (Philadelphia: University of Pennsylvania Press, 2015), 101.

8. On the Progressive Era and related state building, see Robert H. Wiebe, *The Search for Order, 1877–1920* (New York: Hill & Wang, 1967); Richard Hofstadter, *The Age of Reform* (New York: Vintage Books, 1955); Stephen Skowronek, *Building a New American State: The*

Expansion of National Administrative Capacities, 1877–1920 (Cambridge: Cambridge University Press, 1982); Balogh, *Associational State*; Ellis W. Hawley, *The Great War and the Search for a Modern Order: A History of the American People and Their Institutions, 1917–1933*, 2nd ed. (Long Grove, IL: Waveland Press, 1997); and Jurgen Schmandt and James Everett Katz, "The Scientific State: A Theory with Hypotheses," *Science, Technology, and Human Values* 11, no. 1 (1986): 40–52.

9. Quoted in Balogh, *Associational State*, 93.

10. Wiebe, *Search for Order*, 170–71.

11. On conservation efforts in the late nineteenth and early twentieth centuries, see Morton Keller, "Taking Stock," in *Taking Stock: American Government in the Twentieth Century*, ed. Morton Keller and R. Shep Melnick (London: Cambridge University Press, 1999), 1–14; Samuel P. Hays, *Conservation and the Gospel of Efficiency: The Progressive Conservation Movement, 1890–1920* (Pittsburgh: University of Pittsburgh Press, 1999); Robert Higgs, *Crisis and Leviathan: Critical Episodes in the Growth of American Government* (Oakland, CA: The Independent Institute, 2012 [25th anniversary ed.], first published by Oxford, 1987).

12. Daniel P. Carpenter, *The Forging of Bureaucratic Autonomy: Reputations, Networks, and Policy Innovation in Executive Agencies, 1862–1928* (Princeton, NJ: Princeton University Press, 2001), 366; A. Hunter Dupree, *Science in the Federal Government: A History of Policies and Activities* (Baltimore: Johns Hopkins University Press, 1986), 380–81; Ronald G. Walters, ed., *Scientific Authority and Twentieth-Century America* (Baltimore: Johns Hopkins University Press, 1997), 5.

13. Wiebe E. Bijker, Roland Bal, and Ruud Hendriks, *The Paradox of Scientific Authority: The Role of Scientific Advice in Democracies* (Cambridge, MA: MIT Press, 2009), 7. See also Harry Collins and Robert Evans, *Rethinking Expertise* (Chicago: University of Chicago Press, 2007).

14. Yaron Erzahi, *The Descent of Icarus: Science and the Transformation of Contemporary Democracy* (Cambridge, MA: Harvard University Press, 1990), 223; Hale quoted in Walters, "Introduction," *Scientific Authority and Twentieth-Century America*, 1.

15. Nemec, *Ivory Towers*, 11.

16. Balogh, *Associational State*, 116.

17. Walter A. McDougall, . . . *The Heavens and the Earth: A Political History of the Space Age* (Baltimore: Johns Hopkins University Press, 1997) [originally Basic Books, 1985], 75; William E. Akin, *Technocracy and the American Dream: The Technocratic Movement, 1900–1941* (Berkeley: University of California Press, 1977).

18. Brian Balogh, *Chain Reaction: Expert Debate and Public Participation in American Commercial Nuclear Power, 1945–1975* (Cambridge: Cambridge University Press, 1991), 5; David H. DeVorkin, "Who Speaks for Astronomy: How Astronomers Responded to Government Funding after World War II," *Historical Studies in the Physical and Biological Sciences* 31, no. 1 (2000): 55–92.

19. Pestre, "Science, Political Power, and the State"; Carpenter, *Forging of Bureaucratic Autonomy*, 13, 353; Balogh, *Chain Reaction*, 4; Paul R. Josephson, *Resources under Regimes* (Cambridge, MA: Harvard University Press, 2004), 33.

20. Kristine C. Harper, "Meteorology's Struggle for Professional Recognition in the USA (1900–1950)," *Annals of Science* 63, no. 2 (2006): 179–99.

21. Martin V. Melosi, *Precious Commodity: Providing Water for America's Cities* (Pittsburgh: University of Pittsburgh Press, 2011), 9, 12; Hays, *Gospel*, 6–9. While Hays argues that science was important in this decision, Pisani argues that since the science of the early twentieth century was not the same as the science of today, the moral basis of these decisions was more important. See Pisani, *Water and American Government*, 287. The historiography of water in the US West is voluminous. See, for example, Lawrence B. Lee, *Reclaiming the American West: A Historiography and Guide* (Santa Barbara, CA: ABC-Clio Press, 1980); Donald Pisani, *To Reclaim a Divided West: Water, Law, and Public Policy, 1848–1902* (Albuquerque: University of New Mexico Press, 1992); Norris Hundley, Jr., *Water and the West: The Colorado River Compact and the Politics of Water in the American West* (Berkeley: University of California Press, 1975), and *The Great Thirst: Californians and Water, 1770–1990* (Berkeley: University of California Press, 1992); Donald Worster, *Rivers of Empire: Water, Aridity, and the Growth of the American West* (New York: Pantheon Books, 1985).

22. Pisani, *Water and American Government*, xvi.

23. Ibid., xii.

24. Ibid., 2, 279.

25. Ibid., 64; Josephson, *Resources under Regimes*, 35.

26. Nancy Langston, *Forest Dreams, Forest Nightmares: The Paradox of Old Growth in the Inland West* (Seattle: University of Washington Press, 1995), 147.

27. See Josephson, *Resources under Regimes*. For more on state control of US rivers, see, for example, Todd Shallat, *Structures in the Stream* (Austin: University of Texas Press, 1994); and Karen M. O'Neill, *Rivers by Design: State Power and the Origins of US Flood Control* (Durham, NC: Duke University Press, 2006).

28. Langston, *Forest Dreams*, 252–56. See also Christopher McGrory Klyza, *Who Controls Public Lands? Mining, Forestry, and Grazing Policies, 1870–1990* (Chapel Hill: University of North Carolina Press, 1996).

29. Balogh, *Associational State*, 43, 53.

30. Charles E. Closmann, ed., *War and the Environment: Military Destruction in the Modern Age* (College Station: Texas A&M University Press, 2009), 3–4.

31. Langston, *Forest Dreams*, 107–13, 135–36.

32. Susan L. Flader, *Thinking Like a Mountain: Aldo Leopold and the Evolution of an Ecological Attitude toward Deer, Wolves, and Forests* (Madison: University of Wisconsin Press, 1974), 66; Langston, *Forest Dreams*, 240–42.

33. Michael J. Chiarappa, "Overseeing the Family of Whitefishes: The Priorities and Debates of Coregonid Management on America's Great Lakes, 1870–2000," *Environment and History* 11, no. 2 (2005): 163–94; Josephson, *Resources under Regimes*, 48–54; Helen M. Rozwadowski, *The Sea Knows No Boundaries: A Century of Marine Science under ICES* (Copenhagen: International Council for the Exploration of the Sea, with University of Washington Press, Seattle, 2002).

34. Scranton, "Technology, Science and American Innovation," 320–23.

35. Josephson, *Resources under Regimes*, 44–45.

36. Nicolas Rasmussen, "Plant Hormones in War and Peace: Science, Industry, and Government in the Development of Herbicides in 1940s America," *Isis* 92, no. 2 (2001): 315–16.

CHAPTER ONE

1. "Produced a Hard Rain: Experiment with Dyrenforth's Balloons on a Texas Ranch," *Washington Post*, August 20, 1891.

2. See, e.g., National Research Council, *Critical Issues in Weather Modification Research* (Washington, DC: National Academies Press, 2003); James Rodger Fleming, *Fixing the Sky: The Checkered History of Weather and Climate Control* (New York: Columbia University Press, 2010).

3. Mark Fiege, *Irrigated Eden: The Making of an Agricultural Landscape in the American West* (Seattle: University of Washington Press, 2000).

4. For a comprehensive historical account of this early period, see Clark C. Spence, *The Rainmakers: American "Pluviculture" to World War II* (Lincoln: University of Nebraska Press, 1980). George Perkins Marsh, *The Earth as Modified by Human Action* (New York: C. Scribner's Sons, 1874) is a contemporaneous account that discusses the influence of settlement and agriculture on climate. Shorter works focused on the Great Plains include Henry Nash Smith, "Rain Follows the Plow: The Notion of Increased Rainfall for the Great Plains, 1844–1880," *Huntington Library Quarterly* 10, no. 2 (February 1947): 169–93; Walter Kollmorgen, "Rainmakers on the Plains," *Scientific Monthly* 40, no. 2 (February 1935): 146–52; Louise Pound, "Nebraska Rain Lore and Rain Making," *California Folklore Quarterly* 5, no. 2 (April 1946): 129–42.

5. Robert DeC. Ward, "Artificial Rain; A Review of the Subject to the Close of 1889," *American Meteorological Journal* 8, no. 11 (March 1892): 484; Cleveland Abbe, "On the Production of Rain," *Freer's Agricultural Science* 6, no. 7 (July 1892): 297–309. According to Abbe, early researchers/commentators included French physicist and mathematician Dominque-François-Jean Arago and Germans Patricius Heinrich and L. F. Kaemtz. Abbe held that no weather control methods had any practical value, but that "still it is important to briefly summarize the results of experience if only as a means of saving present and future generations from an unnecessary waste of money."; H. C. Russell, "Anniversary Address," Royal Society of New South Wales, May 3, 1882, NOAACL.

6. Ward, "Artificial Rain," 486–87.

7. Russell, "Anniversary Address," 14.

8. On Espy, see James Rodger Fleming, *Meteorology in America, 1800–1870* (Baltimore: Johns Hopkins University Press, 1990).

9. See, for example, Abbe, "On the Production of Rain," 298–99; M. Leschevin, "XXXV. Memoir upon a Process employed in the ci-devant Mâconnais of France, to avert Showers of Hail and to dissipate storms," printed in *Philosophical Magazine* XXVI (October-December 1806, January 1807): 212–18; and Alfred Angot, "Rainfall and Gunfire," *Nature* 99 (August 9, 1917): 467–68. French chemist Charles Le Maout presented his ideas about cannons and ringing bells in his "Météorologie, effets du canon et du son des cloches sur l'atmosphère" (1861).

10. J. K. Laughton, "Can Weather Be Influenced by Artificial Means?" *Nature* 3 (February 16, 1871): 306.

11. Ibid.

12. An excellent overview of technology's role in American life is Thomas Parke Hughes,

American Genesis: A Century of Invention and Technological Enthusiasm, 1870–1970 (Chicago: University of Chicago Press, 2004).

13. Ward, "Artificial Rain," 488. Edward Powers, *War and the Weather, or the Artificial Production of Rain*, rev. ed. (Chicago: Knight & Leonard, 1890).

14. L.E., "IV. Miscellaneous Bibliography; 1. *War and the Weather, or the artificial production of Rain*; by Edward Powers, C.E.," *Silliman's Journal* 2 (1871): 313–14. *Silliman's Journal* was also known as the *American Journal of Science and Arts* and is now the *American Journal of Science*.

15. Prof. I. A. Lapham, "The Great Fires of 1871 in the Northwest," *Journal of the Franklin Institute* 64, no. 1 (July 1972): 46–49.

16. Powers, *War and the Weather*, rev. ed., 3–5.

17. John Wesley Powell, *The Arid Lands*, ed. Wallace Stegner (Lincoln: University of Nebraska Press, 2004 [1878]), 15–16. For more on Powell and the surveys, see Donald Worster, *A River Running West: The Life of John Wesley Powell* (Oxford: Oxford University Press, 2001); and Wallace Stegner, *Beyond the Hundredth Meridian: John Wesley Powell and the Second Opening of the West* (New York: Penguin, 1992 [1954]).

18. Powers, *War and the Weather*, 11.

19. Ibid.

20. Ibid., 200.

21. "The Artificial Production of Rain," *Scientific American* 63, no. 25 (December 20, 1890): 384–85.

22. Ibid.

23. "Rain Makers in the United States," *Scientific American Supplement* 32, no. 84 (October 17, 1891): 131–59.

24. For more on the scientific nature of the US Forest Service and its antecedents, see Paul W. Hirt, *A Conspiracy of Optimism: Management of the National Forests since World War II* (Lincoln: University of Nebraska Press, 1994); and Nancy Langston, *Forest Dreams, Forest Nightmares: The Paradox of Old Growth in the Inland West* (Seattle: University of Washington Press, 1995).

25. General Robert G. Dyrenforth and Professor Simon Newcomb, LL.D., "Can We Make It Rain?" *North American Review* 153, no. 419 (October 1891): 393.

26. Dyrenforth and Newcomb, "Can We Make It Rain?" 395–98.

27. George E. Curtis, "Rain-making in Texas," *Nature* 44, no. 1147 (October 22, 1891): 594.

28. Ibid.

29. Ibid.

30. George E. Curtis, "Artificial Rain-Making," *Western Christian Advocate* (Cincinnati, OH), September 24, 1891. Examples of dueling articles/letters to the editor about the efficacy of these rainmaking efforts written at the time of the experiments include H. A. Hazen, "The Rain-Makers," *Science* 18, no. 447 (August 8, 1891): 122–23, "Rain-Making," *Science* 18 (454) (October 16, 1891): 219, and "Battles and Rain," *Science* 18, no. 457 (November 6, 1891): 264; Edward Powers, Letter to the Editor, "Rain-Making," *Science* 18, no. 453 (October 9, 1891): 205–6, and "Rain-Making," *Science* 18, no. 456 (October 30, 1891): 249.

31. George E. Curtis, "The Rain Making Experiments in Texas; An Interesting Statement from the Meteorologist of the Expedition," *Inventive Age* 3 (December 29, 1891): 2.

32. See, e.g., Thomas Parke Hughes, *Changing Attitudes toward American Technology* (New York: Harper & Row, 1975).

33. The point about avoiding waste is made by Langston, *Forest Dreams*; Donald J. Pisani, *Water and American Government: The Reclamation Bureau, National Water Policy, and the West, 1902–1935* (Berkeley: University of California Press, 2002) and Paul R. Josephson, *Resources under Regimes* (Cambridge, MA: Harvard University Press, 2004.

34. Dyrenforth and Newcomb, "Can We Make It Rain?" 398–99.

35. Ibid., 402. Newcomb was referring to the Charles Lamb essay "A Dissertation upon Roast Pig," first published in *London Magazine*, September 1882, and later appearing in the collection of his work titled *The Essays of Elia*.

36. On experts, expertise, and the state, see, for example, Wiebe E. Bijker, Roland Bal, and Ruud Hendriks, *The Paradox of Scientific Authority: The Role of Scientific Advice in Democracies* (Cambridge, MA: MIT Press, 2009); Yaron Ezrahi, *The Descent of Icarus: Science and the Transformation of Contemporary Democracy* (Cambridge, MA: Harvard University Press, 1990); Mark R. Nemec, *Ivory Towers and Nationalistic Minds: Universities, Leadership, and the Development of the American State* (Ann Arbor: University of Michigan Press, 2006); Brian Balogh, *The Associational State: American Governance in the Twentieth Century* (Philadelphia: University of Pennsylvania Press, 2015), and *Chain Reaction: Expert Debate and Public Participation in American Commercial Nuclear Power, 1945–1975* (Cambridge: Cambridge University Press, 1991).

37. Dyrenforth and Newcomb, "Can We Make It Rain?" 404.

38. "Physicists in Doubt: Their Views on the Experiments for Inducing Rainfall," *Washington Post*, August 21, 1891. See, e.g., J. Aitken, "On Dust, Fogs, and Clouds," *Proceedings of the Royal Society of Edinburgh* 11, no. 108 (1880–81): 14–18, 122–26.

39. Walter J. Grace, "Notes and Comments on 'Harnessing the Rain-Cloud,'" *North American Review* 153, no. 147 (August 1891): 252–53.

40. "Rain-making Experiments," *Washington Post*, August 18, 1891.

41. "'Work of the Rain-Makers': How It Is Caused to Fall on the Just and the Unjust," *Boston Sunday Herald*, September 13, 1891.

42. "Lieut. Ellis' Experiments," *Washington Post*, October 2, 1891.

43. "Rain Makers in the United States," *Scientific American Supplement* 32, no. 824 (October 17, 1891): 13159.

44. "The Cloud Compellers and the Press," *Scientific American Supplement* 32, no. 824 (October 17, 1891): 13161.

45. Chief of the Division of Forestry to Secretary Rusk, quoted in Robert DeC. Ward, "Artificial Rain," 492.

46. William Morris Davis, "The Theories of Artificial and Natural Rainfall," *American Meteorological Journal* 8, no. 11 (March 1892): 493.

47. George E. Curtis, "The Facts about Rain-Making," *Engineering Magazine* 3 (July 1892): 548–49.

48. Ibid., 549.

49. Ibid., 550–51.

50. "The Recent Rainfall Experiments in San Antonio, Texas," *American Meteorological Journal* 9, no. 10 (February 1893): 459; Secretary Morton's standard reply to inquiries about artificial rainfall experiments, received at the Weather Bureau Library on May 15, 1894, NOAACL.

51. "He Gets Rain through a Hole in the Roof," *Washington Star*, September 2, 1891; "A Man Who Can Make Rain," *Augusta* (Georgia) *Chronicle*, ca. July 1891, NOAACL.

52. B. S. Pague, "Artificially Increasing Rainfall," ca. 1899, NOAACL.

53. C. W. Post, "Making Rain While the Sun Shines," *Harper's Weekly* 56, part 1 (February 24, 1912): 8.

54. Ibid.

55. "'Making' Rain in Battle Creek," *Scientific American* 107, no. 6 (August 10, 1912): 10.

56. "Rain after Battles," *Scientific American* 111, no. 17 (October 24, 1914): 330.

57. "Rain-making," *Times of London* editorial, issue 42856 (October 20, 1921): 11; "The Rain-Maker: Fighting Drought with Chemicals," *Illustrated London News*, NOAACL.

58. A. H. Palmer, "Professional 'Rain-Makers,'" *BAMS* 1 (1920): 47–48. The American Geophysical Union and the International Union of Geodesy and Geophysics were also formed in 1919. Their efforts to bring scientific expertise to government officials made them part of what Brian Balogh calls the "associational state." See Balogh, *Associational State*.

59. "More about the 'Rain-Maker,'" *BAMS* 1 (1920): 80–82.

60. "Let Us Change the Ocean Currents and Our Climate," *BAMS* 2 (1921): 111–12.

61. William R. Blair to W. F. G. Swann, July 7, 1921, Box 1/Rainmaking 3, Chaffee.

62. Swann to Blair, July 19, 1921, Box 1/Rainmaking 3, Chaffee.

63. E. L. Chaffee to L. Frances Warren, August 15, 1921, Box 1/Rainmaking 3, Chaffee.

64. For the history of American aviation in the early twentieth century, see, e.g., Roger E. Bilstein, *Flight in America, 1900–1983: From the Wrights to the Astronauts* (Baltimore: Johns Hopkins University Press, 1984).

65. Wilder D. Bancroft to Chaffee, November 18, 1921; Warren to Chaffee, March 25, 1922, Box 1/Rainmaking 3, Chaffee.

66. "How Electric Sand Can Dispel Clouds Told by Discoverer," *NYT*, February 15, 1923, 1.

67. J. P. Staples to Chaffee, May 16, 1922, Box 1/Rainmaking 4, Chaffee.

68. John D. Price, LT, USN, statement witnessed by J. R. Hegman, June 30, 1922, Box 1/Rainmaking 4, Chaffee.

69. "Rain Making: Successful Experiments in America," *Times of London*, February 13, 1923, issue 43264, 11, column F; "How Electric Sand," *NYT*, February 15, 1923, 1.

70. "How Electric Sand," *NYT*, February 15, 1923, 1.

71. Ibid.

72. Major Frank to Chaffee, February 19, 1923, Box 1/Rainmaking 4, Chaffee.

73. "How Electric Sand," *NYT*, February 15, 1923, 1.

74. Ibid.

75. Warren to Knight and Chaffee, September 1, 1923, Box 1/Rainmaking 4, Chaffee.

76. Knight to Warren, September 4, 1923, Box 1/Rainmaking 4, Chaffee.

77. Major E. A. Lohman to R. O. Chaffee, September 27, 1924, Box 1/Rainmaking 1, Chaffee.

78. E. L. Chaffee to Major Lohman, October 28, 1924, Box 1/Rainmaking 2, Chaffee.

79. "End Fog, Bring Rain by Shooting Clouds," *NYT*, October 30, 1924, 1.

80. James C. Young, "Rain at Man's Will Is Declared Possible," *NYT*, November 9, 1924, XX5–XX6. Italics added. This is the first instance of the term "weather control" appearing in the *New York Times*.

81. Dust Bowl conditions dried up rainmaking, but not dam building, which continued

apace during the Great Depression. Most of those dams, however, were not built to meet demand for irrigation or hydroelectric power at the time, but to provide employment. The same was true for those working for the Civilian Conservation Corps, which was more about jobs than about conservation. The payoff for the nation came in World War II. See Neil M. Maher, *Nature's New Deal* (Oxford: Oxford University Press, 2008).

82. Sir Napier Shaw, *The Air and Its Ways* (Cambridge: University Press, 1923), 216.

83. Ibid., 225.

84. Alexander McAdie, "Prankish Clouds and Patient Men," *NYT*, April 22, 1923, BR9.

85. Alexander McAdie, *Making the Weather* (New York: MacMillan, 1923).

86. W. J. Humphreys, *Rain Making and Other Weather Vagaries* (Baltimore: Williams & Wilkins, 1926), 3, 28–94.

87. Sir Oliver Lodge, "Electrical Precipitation: A Lecture Delivered before the Institute of Physics" (London: Oxford University Press, 1925).

88. Richard D. Coons, Robert C. Gentry, and Ross Gunn, "First Partial Report on the Artificial Production of Precipitation: Stratiform Clouds, Ohio, 1948," Department of Commerce, Research Paper No. 30, Washington DC, Box 38/1, Langmuir.

89. A. Wegener, *Thermodynamik der Atmosphäre* (J. Barth, 1911).

90. Tor Bergeron, "Über die dreidimensional verknüpfende Wetteranalyse," *Geofysiske Publikasjoner* 5 (1928): 1–111.

91. T. Bergeron, "On the Physics of Clouds and Precipitation," Proc. Vᵉ Assemblée Générale de l'Union Géodésique et Geophysique Internationale, Lisbon, Portugal, 1935, International Union of Geodesy and Geophysics, 156–80; Tor Bergeron, "Some Autobiographic Notes in Connection with the Ice Nucleus Theory of Precipitation Release," *BAMS* 59 (1978): 390–92. See also Arnt Eliassen, "Tor Bergeron 1891–1977," *BAMS* 59 (1978): 387–89, and Gosta H. Liljequist, "Tor Bergeron: A Biography," *Pure and Applied Geophysics* 119 (1981): 409–42.

92. W. Findeisen, "Kolloid-meteorologische Vorgänge bei Neiderschlags-bildung," *Meteorologische Zeitschrift* 55 (1938): 121–33.

93. See, e.g., Roscoe R. Braham, Jr., "Formation of Rain: A Historical Perspective," in *Historical Essays on Meteorology, 1919–1996; The Diamond Anniversary Volume of the American Meteorological Society*, ed. James Rodger Fleming (Boston: American Meteorological Society, 1996); P. V. Hobbs, "The Scientific Basis, Techniques, and Results of Cloud Modification," in *Weather Modification: Science and Public Policy,* ed. Robert G. Fleagle, 30–42 (Seattle: University of Washington Press, 1969).

94. "Summary of Results on the Investigation of the Practicability of Dissipating Natural Fog by Means of Hygroscopic Material for the Year 1935–36 (As of July 1936)," August 12, 1936; Edward L. Bowles to Chief, Bureau of Aeronautics (Navy), August 12, 1936, Box 4/14, Houghton; Robert G. Stone, "Controlled Weather," *BAMS* 15 (1934): 205.

95. Stone, "Controlled Weather," 205, quoting a story from the *Christian Science Monitor* of August 4, 1934.

CHAPTER TWO

1. "Sarnoff Predicts Weather Control and Delivery of the Mail by Radio," *NYT*, October 1, 1946, 1.

2. See, e.g., Lisa Rosner, ed., *The Technological Fix: How People Use Technology to Create and Solve Problems* (New York: Routledge, 2004). Nuclear physicist Alvin Weinberg, *Reflections on Big Science* (Oxford: Pergamon Press, 1967), was convinced that given enough time, technology could fix whatever ailed us.

3. Justification Memorandum, PD #EN1-22/00028, the Institute for Advanced Study, June 6, 1946, Box 15/6, von Neumann. For more on the development of numerical weather prediction, see Kristine C. Harper, *Weather by the Numbers: The Genesis of Modern Meteorology* (Cambridge, MA: MIT Press, 2008).

4. For a discussion of the methodological differences between lab and field sciences, see Robert E. Kohler, *Landscapes and Labscapes: Exploring the Lab-Field Border in Biology* (Chicago: University of Chicago Press, 2002). Foresters also found out that what worked on trees in the laboratory did not always work in the forest. See Nancy Langston, *Forest Dreams, Forest Nightmares: The Paradox of Old Growth in the Inland West* (Seattle: University of Washington Press, 1995), 135-36.

5. For an introductory overview of Cold War–era views of science and technology, see Audra J. Wolfe, *Competing with the Soviets: Science, Technology, and the State in Cold War America* (Baltimore: Johns Hopkins University Press, 2103).

6. Opaque, granular rime ice forms when supercooled (below freezing temperature) water freezes rapidly. Transparent clear ice is the greater threat since it sticks firmly to the aircraft, significantly reducing lift.

7. E. F. Black to George F. Doriot, Business Administration School, Harvard, May 1, 1947 [Confidential], Box 469/9, RDB.

8. Irving Langmuir et al., "Meteorological Research" (Schenectady, NY: General Electric Research Laboratory, 1947). See also Kristine C. Harper, "Climate Control: United States Weather Modification in the Cold War and Beyond," *Endeavour* 32, no. 1 (2008): 20-26.

9. Langmuir et al., "Meteorological Research."

10. Black to Doriot, May 1, 1947, Box 469/9, RDB.

11. Langmuir et al., "Meteorological Research."

12. Ibid.

13. Black to Doriot, May 1, 1947, Box 469/9, RDB.

14. Langmuir et al., "Meteorological Research." Bernard Vonnegut was novelist Kurt Vonnegut's brother.

15. Irving Langmuir, "List of Problems with Ice Nucleation within Clouds," December 12, 1946, Box 11/4, Langmuir.

16. Dan Kevles, "Cold War and Hot Physics: Science, Security, and the American State, 1945-1946," *Historical Studies in the Physical and Biological Sciences* 20, no. 2 (1990): 246-47. For another take on the RDB, see Deborah Jean Warner, "Political Geodesy: The Army, the Air Force, and the World Geodetic System of 1960," *Annals of Science* 59, no. 4 (2002): 363-89.

17. L. V. Berkner to Planning Division, JRDB (Joint Research and Development Board; later just Research and Development Board [RDB]), November 22, 1946; D. B. Langmuir to Chair, Committee on Geophysical Sciences [SECRET], February 21, 1947, Box 469/9, RDB.

18. Roland F. Beers to D. B. Langmuir, March 5, 1947, Box 469/9, RDB. For more on Rossby's influence in US government circles, see Harper, *Weather by the Numbers*.

19. Carl-Gustav Rossby to the Committee on Geophysical Sciences, April 7, 1947 [CONFI-DENTIAL], Box 469/9, RDB.

20. See, e.g., Nikolai Krementsov, *The Cure: A Story of Cancer and Politics from the Annals of the Cold War* (Chicago: University of Chicago Press, 2002), on the KR story.

21. Rossby to Committee on Geophysical Sciences, April 7, 1947 [CONFIDENTIAL], Box 469/9, RDB. Federal attempts to regulate weather control are addressed in chap. 3. In September 1947, the War Department became the Department of the Army, and the Department of the Air Force was established.

22. D. B. Langmuir to Col. Haugen, Lt. Col. Black, and Capt. Morris, July 14, 1947, Box 459/9, RDB.

23. L. T. Morse to Executive Council, August 11, 1947, Box 459/9, RDB.

24. Frances L. Whedon to Helmut Landsberg, August 20, 1946 [CONFIDENTIAL], Box 459/9, RDB.

25. Policy on Cloud Modification Studies, November 25, 1947, Box 228/Cloud Modification Studies 1947–48/2; Vannevar Bush to James Forrestal, January 15, 1948 [CONFIDENTIAL]; Helmut Landsberg to Francis Reichelderfer, February 10, 1948 [CONFIDENTIAL]; L. R. Hafstad to Joint Staff, February 5, 1948 [SECRET], Box 459/9; Whedon memo for the Secretariat, Panel on Meteorology, RDB, May 7, 1948, Box 485/Info on Budgets, RDB.

26. Hafstad to Joint Staff, February 5, 1948 [SECRET], Box 459/9, RDB.

27. "Army and GE to 'Make Weather,'" *NYT*, March 14, 1947, 25.

28. Landsberg to General H. S. Aurand, March 21, 1947; Landsberg memo, March 31, 1947; Col. James S. Willis to Landsberg, April 7, 1947, Box 469/9, RDB. The military could use nature in other ways as well. See Jacob Darwin Hamblin, *Arming Mother Nature: The Birth of Catastrophic Environmentalism* (Oxford: Oxford University Press, 2013).

29. Earl G. Droessler to Landsberg, April 21, 1947, Box 469/9, RDB.

30. Ibid. Officially, the "snow project" was Army Project 2001–7 and Navy NR 082016.

31. Irving Langmuir et al., "Meteorological Research."

32. Ibid.

33. Ibid.

34. "Finds a New Way to Produce Rain," *NYT*, November 18, 1947, 31; Barrington S. Havens et al., "Early History of Cloud Seeding" (Socorro: Langmuir Laboratory, New Mexico Institute of Mining and Technology, 1978), 27.

35. Havens et al., "Early History," 28. The Thunderstorm Project was conducted by the University of Chicago.

36. "House Bill to Ask Study of Weather Control by US," *NYT*, September 8, 1947, 13; H.R. 4582, November 25, 1947, Box 459/9, RDB.

37. D. E. Chambers to Committee on Interstate and Foreign Commerce, House of Representatives, March 16, 1948, Box 459/9, RDB.

38. Clinton P. Anderson to Hon. Charles A. Wolverton, March 17, 1948, Box 522/Working Drafts, Anderson. Based on Richard Allen Baker's *Conservation Politics: The Senate Career of Clinton P. Anderson* (Albuquerque: University of New Mexico Press, 1985), Anderson's interest in weather control stemmed from his focus on water for agriculture, a major concern in his very dry state.

39. Statement by B. G. Shaw, March 18, 1948, Box 581/1, RDB.

40. Forest Service Statement on H.R. 4582, March 17, 1948, Box 459/9, RDB.

41. Ross Gunn to Francis Reichelderfer, May 27, 1948, Box 459/9, RDB. The Navy contributed ground-based meteorological equipment.

42. Ibid.

43. Ibid. For example, a circular cluster of thunderstorms (known as a mesoscale convective cluster) may develop new cumulonimbus cells as the older ones die out; the downdraft of the dying cells feeds the updraft of the new ones.

44. Richard D. Coons, Robert C. Gentry, and Ross Gunn, "First Partial Report on the Artificial Production of Precipitation: Stratiform Clouds, Ohio, 1948," Department of Commerce, Research Paper No. 30, Washington DC, Box 38/1, Langmuir.

45. "Air Force Drops Rain-Making Tests," (AP Story) *Schenectady Gazette*, November 23, 1948, Box 101/Clips-1, Langmuir.

46. "The Nation: Forecast: Dry Spell," *NYT*, November 28, 1948, E1; "Hasty Conclusion," *Schenectady Gazette,* December 1, 1948, Box 1/Clips-1, Langmuir.

47. Joint AF-USWB Release No. 94, November 28, 1948, Box 6/Records of Cloud Physics Ad Hoc Operational Evaluation Committee, 1948–1950, USWB.

48. "Rain-making Advanced," *NYT*, 30 November 1948, 1.

49. Reichelderfer to Group Captain H. N. Warren, 30 December 1948, Entry 124, Box 10, USWB.

50. Richard S. Coons, Earl L. Jones, and Ross Gunn, "Second Partial Report on the Artificial Production of Precipitation," 1948, Box 38/11, Langmuir.

51. Memorandum Report, US Air Force Air Materiel Command, Wright Field, Dayton, Ohio, Serial No. AWNW 9-4, Subject: Preliminary Report on Phase I, II—Cloud Physics Project, January 3, 1949 [RESTRICTED], Box 109, Langmuir MSS.

52. Michael Ference to Robert L. Sorey, January 14, 1949, Box 6/Cloud Physics, 1948–1950, USWB.

53. Ibid.

54. Irving Langmuir, "The Growth of Particles in Smokes and Clouds and the Production of Snow from Supercooled Clouds," *Proceedings of the American Philosophical Society* 92, no. 3 (July 1948): 180.

55. Ibid., 170–82.

56. Ibid., 183.

57. Ibid., 183–84.

58. Irving Langmuir, "Outline of Progress in the Evaluation of Cloud Modification Techniques," 1948, p. 26, Box 83/1948, Langmuir.

59. Ibid., 19.

60. Ibid., 29.

61. See Kohler, *Landscapes and Labscapes.*

62. Irving Langmuir, "Large Scale Seeding of Stratus and Cumulus Clouds with Dry Ice," January 25, 1949, Box 90/149, Langmuir.

63. Francis Reichelderfer, "What the Aeronautical Sciences are Doing about the Weather." Abstract of paper given at the 1949 Annual AMS Meeting, January 25, 1949, Box 1/1, Reichelderfer.

64. Reichelderfer—Memo for Desk Reference, January 28, 1949, Entry 124, Box 10; Reichelderfer to Chair, Ad Hoc Operational Evaluation Committee of the Cloud Physics Project, Box 6/Cloud Physics 1948–1950; Reichelderfer to J. A. West, February 3, 1949, Box 10, USWB.

65. Reichelderfer to Gunn, February 4, 1949, Entry 124, Box 10, USWB.

66. Reichelderfer, "Recent Research in Cloud Physics with Special Reference to Production of Rain," Cosmos Club lecture, February 7, 1949, Box 1/4, Reichelderfer.

67. Reichelderfer to Delbert Little, May 12, 1949, Entry 124, Box 10, USWB.

68. Panel of the Atmosphere—1949 Technological Estimate [CONFIDENTIAL], April 14, 1949, Box 171/2, RDB.

69. William Lewis to Langmuir, June 23, 1949, Box 12/4, Langmuir.

70. Proposed Plan for Operational Evaluation of Cloud Seeding Techniques July 1, 1949–June 30, 1950, Box 6/Cloud Physics 1948–1950, USWB.

71. Ibid.

72. Major Robert L. Sorey to Director, Plans and Operations, USAF; Chief, Weather Bureau; and Admin Aide to the Chief of Naval Operations [CONFIDENTIAL], July 7, 1949, Box 6/Cloud Physics 1948–1950, USWB.

73. Major Robert L. Sorey to Chief, Weather Bureau and Commanding General, Air Weather Service [CONFIDENTIAL], July 7, 1949, Box 6/Cloud Physics 1948–1950, USWB; Minutes of the Panel on the Atmosphere Working Group, August 12, 1949, Box 458/3, RDB.

74. Irving Langmuir, "Study of Periodicity in Rainfall due to Silver Iodide Seeding in New Mexico," September 29, 1950, Box 83/1950–1, Langmuir.

75. Ference to Landsberg, March 8, 1950, Box 459/8, RDB.

76. Hurd C. Willett to Landsberg, March 10, 1950, Box 459/8, RDB.

77. Gardner Emmons, George P. Wadsworth, Hurd C. Willett, and Bernhard Haurwitz to Landsberg, March 30, 1950, Box 459/8, RDB.

78. Landsberg to Ference, April 14, 1950; Emmons et al. to Landsberg, March 30, 1950, Box 459/8, RDB.

79. Landsberg to Ference, April 14, 1950; Cary J. King, Jr., to Executive Director, Committee on Geophysics and Geography, June 2, 1950, Box 459/8, RDB.

80. Roger Hammond to Langmuir, June 9, 1950, Box 6/1955, Langmuir. For a brief summary of flood conditions, see "SOS! Canadian Disasters," Library and Archives Canada, accessed March 15, 2013, http://www.collectionscanada.gc.ca/sos.

81. "Easy to Make Rain, Langmuir Asserts," *NYT* July 16, 1950, 56; Irving Langmuir, "Control of Precipitation from Cumulus Clouds by Various Seeding Techniques," *Science* 112, no. 2898 (July 14, 1950): 35–41.

82. Ferguson Hall, Gardner Emmons, Bernhard Haurwitz, George P. Wadsworth and Hurd C. Willett, "Dr. Langmuir's Article on Precipitation Control," *Science* New Series 113, no. 2929 (February 16, 1951): 189–192.

83. William Webster to Landsberg, July 7, 1950 [RESTRICTED], Box 459/8, RDB.

84. Secretary of Commerce to Raymond P. Whearty, July 7, 1950 (circa), Entry 120, Box 1, USWB.

85. Panel on the Atmosphere, 1950 Technical Estimate Draft 1, April 3, 1950, Box 171/2, RDB.

86. Panel on the Atmosphere 1950 Technical Estimate (Final) [CONFIDENTIAL], June 14, 1950, Box 171/2, RDB.

87. Irving Langmuir, "Study of Periodicity in Rainfall due to Silver Iodide Seeding in New Mexico," September 29, 1950, Box 83/1950–1, Langmuir.

88. John Pfeiffer, "Scientist of Light and Weather," *NYT*, January 28, 1951, SM6.

89. Robert C. Cowan, "Langmuir Ties US Weather to Cloud Seeding," *Christian Science Monitor*, October 13, 1950, Box 498/Background Material Weather Control Act, Anderson.

90. Irving Langmuir, "Widespread Modifications of Synoptic Weather Conditions Induced by Localized Silver Iodide Seeding," GE Research Lab, Schenectady, NY, circa 1951, Box 84/1951–1, Langmuir.

91. Langmuir to Reichelderfer, December 1, 1950, Box 6/1950, Langmuir.

92. "Weather Control Called 'Weapon,'" (UPI) *NYT*, December 10, 1950, 68.

93. "GE Aids Rainmaking," *NYT*, December 28, 1950, 25.

94. "Langmuir Quits Post to Push Rainmaking," *NYT*, March 2, 1950, 40.

PART TWO

1. Aaron L. Friedberg, *In the Shadow of the Garrison State: America's Anti-Statism and Its Cold War Grand Strategy* (Princeton, NJ: Princeton University Press, 2000), 3–4, Tilly quoted on page 3.

2. Jurgen Schmandt and James Everett Katz, "The Scientific State: A Theory with Hypotheses," *Science, Technology, and Human Values* 11, no. 1 (1986): 40–52.

3. Brian Balogh, *The Associational State: American Governance in the Twentieth Century* (Philadelphia: University of Pennsylvania Press, 2015), 91.

4. Ibid.; Paul Forman, "Behind Quantum Electronics: National Security as Basis for Physical Research in the United States," *Historical Studies in the Physical and Biological Sciences* 18, no. 1 (1987): 149–229; Scott Frickel and Kelly Moore, eds., *The New Political Sociology of Science: Institutions, Networks, and Power* (Madison: University of Wisconsin Press, 2006); and David M. Hart, *Forged Consensus: Science, Technology, and Economic Policy in the United States, 1921–1953* (Princeton, NJ: Princeton University Press, 1998). Historical treatments of the national security state (itself a contested term), and how deterrence and mobilization influenced by science and technology played out in the United States, are extensive. See, e.g., John L. Gaddis, *Strategies of Containment* (New York: Oxford University Press, 1982); Michael J. Hogan, *Cross of Iron: Harry S. Truman and the Origins of the National Security State, 1945–1954* (Cambridge: Cambridge University Press, 1998); Samuel P. Huntington, *The Common Defense: Strategic Programs in National Politics* (New York: Columbia University Press, 1961); Fred Kaplan, *The Wizards of Armageddon* (New York: Simon & Schuster, 1983); Melvyn P. Leffler, *A Preponderance of Power: National Security, the Truman Administration, and the Cold War* (Stanford, CA: Stanford University Press, 1992); Richard M. Merelman, "Technological Cultures and Liberal Democracy in the United States," *Science, Technology, and Human Values* 25, no. 2 (2000): 167–94; Walton S. Moody, *Building a Strategic Air Force* (Washington, DC: Center for Air Force History, 1996); Richard Rhodes, *Dark Sun: The Making of the Hydrogen Bomb* (New York: Simon & Schuster, 1995); Michael S. Sherry, *The Rise of American Air Power* (New Haven, CT: Yale University Press, 1987); and Daniel Yergin, *Shattered Peace: The Origins of the Cold War and the National Security State* (Boston: Houghton, Mifflin, 1977).

5. Vannevar Bush, *Science—the Endless Frontier* (July 1945) (Washington, DC: National

Science Foundation, 1960), 6–12, and Hart, *Forged Consensus*. Caltech President Lee DuBridge agreed with Bush, arguing that in the past the United States had used science from abroad instead of creating it domestically. See Matthew Shindell, "From the End of the World to the Age of the Earth: The Cold War Development of Isotope Geochemistry at the University of Chicago and Caltech," in *Science and Technology in the Global Cold War*, ed. Naomi Oreskes and John Krige (Cambridge, MA: MIT Press, 2014), 107–39.

6. Stuart W. Leslie, *The Cold War and American Science: The Military-Industrial-Academic Complex at MIT and Stanford* (New York: Columbia University Press, 1993), 1, 11. See Ronald E. Doel, "Constituting the Postwar Earth Sciences: The Military's Influence on the Environmental Sciences in the USA after 1945," *Social Studies of Science* 33, no. 5 (2003): 635–66.

7. Bruce Kuklick, *Blind Oracles: Intellectuals and War from Kennan to Kissinger* (Princeton, NJ: Princeton University Press, 2006), 14. The National Institutes of Health were also a major funder, but NIH and NSF paled compared to the Atomic Energy Commission and military-related agencies. Both sides of the US/USSR rivalry capitalized on it to obtain funding for science related to national security. See, e.g., Mark Solovey, "Introduction: Science and the State during the Cold War: Blurred Boundaries and a Contested Legacy," *Social Studies of Science* 31, no. 2 (2001): 165–70; Philip Scranton, "Technology, Science and American Innovation," *Business History* 48, no. 3 (2006); Everett Mendelsohn, Merritt Roe Smith, and Peter Weingart, eds., *Science, Technology and the Military [Sociology of the Sciences, A Yearbook, 12, Part I]* (Dordrecht: Kluwer Academic Publishers, 1988); Robert Gilpin and Christopher Wright, eds., *Scientists and National Policy-Making* (New York: Columbia University Press, 1964); Patrick J. McGrath, *Scientists, Business, and the State, 1890–1960* (Chapel Hill: University of North Carolina Press, 2002); Lawrence Busch, *The Eclipse of Morality: Science, State and Market* (New York: Aldine de Gruyter, 2000); James H. Capshaw and Karen A. Rader, "Big Science: Price to the Present," in *Science after '40*, ed. Arnold Thackray, *Osiris* 7 (1992): 3–25; Roger L. Geiger, "Science, Universities, and the National Defense," in *Science After '40*, ed. Arnold Thackray, *Osiris* 7 (1992): 26–48; Chandra Mukerji, *A Fragile Power: Scientists and the State* (Princeton, NJ: Princeton University Press, 1989). On the use of rhetoric to keep money flowing, see David J. Tietge, *Flash Effect: Science and the Rhetorical Origins of Cold War America* (Athens: Ohio University Press, 2002).

8. Naomi Oreskes, "Introduction," in *Science and Technology in the Global Cold War*, ed. Naomi Oreskes and John Krige (Cambridge, MA: MIT Press, 2014), 3. Literature on the state's involvement with science in the United States is voluminous and includes J. Merton England, *A Patron for Pure Science: The National Science Foundation's Formative Years, 1945–1957* (Washington, DC: National Science Foundation, 1982); Daniel Lee Kleinman, *Politics on the Endless Frontier: Postwar Research Policy in the United States* (Durham, NC: Duke University Press, 1995); Walter A. McDougall, . . . *The Heavens and the Earth: A Political History of the Space Age* (Baltimore: Johns Hopkins University Press, 1997) [originally Basic Books, 1985]; Jessica Wang, *American Science in the Age of Anxiety: Scientists, Anticommunism, and the Cold War* (Chapel Hill: University of North Carolina Press, 1999); David B. Resnick, *Playing Politics with Science: Balancing Scientific Independence and Government Oversight* (New York: Oxford University Press, 2009); Forman, "Behind Quantum Electronics"; Daniel J. Kevles, "K_1S_2: Korea, Science, and the State," in *Big Science: The Growth of Large-Scale Research*, ed. Peter Galison and Bruce Hevly (Stanford, CA: Stanford University Press, 1992), 312–54; Larry

Owens, "Science in the United States," in *Science in the Twentieth Century*, ed. John Krige and Dominique Pestre (Amsterdam: Harwood Academic Publishers, 1997), 821–37; Asif Siddiqi, "Fighting Each Other: The N-1, Soviet Big Science and the Cold War at Home," in *Science and Technology in the Global Cold War*, ed. Naomi Oreskes and John Krige (Cambridge, MA: MIT Press, 2014), 189–225; Ronald E. Doel, "The Earth Sciences and Geophysics," in *Science in the Twentieth Century*, ed. John Krige and Dominique Pestre (Amsterdam: Harwood Academic Publishers, 1997), 391–416; John Krige and Dominique Pestre, eds., *Science in the Twentieth Century* (Amsterdam: Harwood Academic Publishers, 1997); Stanley Aronowitz, *Science as Power: Discourse and Ideology in Modern Society* (Houndmills, UK: Macmillan, 1988); Michael Fortun and Sylvan S. Schweber, "Scientists and the State: The Legacy of World War II," in *Trends in the Historiography of Science*, ed. Kostas et al. (Dordrecht: Kluwer Academic Publishers, 1994), 327–54; Michael Aaron Dennis, "Reconstructing Sociotechnical Order: Vannevar Bush and US Science Policy," in *States of Knowledge: The Co-production of Science and Social Order*, ed. Sheila Jasanoff (London: Routledge, 2004), 225–53; and Sheila Jasanoff, *Designs on Nature: Science and Democracy in Europe and the United States* (Princeton, NJ: Princeton University Press, 2005).

9. On bureaucratic autonomy, see Daniel P. Carpenter, *The Forging of Bureaucratic Autonomy: Reputations, Networks, and Policy Innovation in Executive Agencies, 1862–1928* (Princeton, NJ: Princeton University Press, 2001); Stephen D. Krasner, *Defending the National Interest: Raw Materials Investment and U.S. Foreign Policy* (Princeton, NJ: Princeton University Press, 1978); and Peter B. Evans, Dietrich Rueschemeyer, and Theda Skocpol, eds., *Bringing the State Back In* (Cambridge: Cambridge University Press, 1985).

10. Harry Collins and Robert Evans, *Rethinking Expertise* (Chicago: University of Chicago Press, 2007), 14–51, 143. Additional works on the role of expert advice as it relates to science and politics include Stephen Hilgartner, *Science on Stage: Expert Advice as Public Drama* (Stanford, CA: Stanford University Press, 2000); Wiebe E. Bijker, Roland Bal, and Ruud Hendriks, *The Paradox of Scientific Authority: The Role of Scientific Advice in Democracies* (Cambridge, MA: MIT Press, 2009); Bruce Bimber, *The Politics of Expertise in Congress: The Rise and Fall of the Office of Technology Assessment* (Albany: State University of New York Press, 1996); Brian Balogh, *Chain Reaction: Expert Debate and Public Participation in American Commercial Nuclear Power, 1945–1975* (Cambridge: Cambridge University Press, 1991); Harry Collins, *Are We All Scientific Experts Now?* (Cambridge: Polity, 2014); and Ronald G. Walters, ed., *Scientific Authority and Twentieth-Century America* (Baltimore: Johns Hopkins University Press, 1997).

CHAPTER THREE

1. Quoted in Frederick C. Othman, "Rain of Power," *NY World Telegram*, November 12, 1950, Box 498/Weather Control Inquiries, Anderson.

2. Louis Reid, "Science Sees Made-to-Order Weather," *New York Journal-American*, April 23, 1950, 4-L.

3. William A. Ward to Clinton P. Anderson, September 17, 1950, Box 498/Weather Control Inquiries, Anderson.

4. Triviz to Anderson, September 17, 1950; Memo to Anderson, September 18, 1950, Box 522/Working Drafts, Anderson.

5. Triviz to Ernest S. Griffith, September 20, 1950, Box 522/Working Drafts, Anderson.

6. American Law Section, Legislative Reference Section, to Anderson, September 28, 1950, Box 522/Working Drafts, Anderson.

7. Ibid.

8. Reichelderfer to Anderson, October 9, 1950, Box 522/Working Drafts, Anderson.

9. Triviz to Anderson, October 10, 1950, Box 522/Working Drafts, Anderson.

10. "State Group asks for Federal Control of Cloud Seeding," *Albuquerque Journal*, October 14, 1950, Box 498/Background Material Weather Control Act, Anderson.

11. Drew Pearson, "Potentials of Artificial Rain-Making," *Washington Post*, October 27, 1950, 13C.

12. Triviz to Anderson, November 1, 1950, Box 522/Working Drafts, Anderson.

13. Reichelderfer to Anderson, November 1, 1950; Triviz memo, November 3, 1950, Box 522/Working Drafts, Anderson.

14. Workman to Anderson, November 6, 1950, Box 499/Correspondence—Comments on Shaping Weather Control Legislation, Anderson.

15. John E. Pernice to Triviz, November 20, 1950, Box 499/Comments, Anderson.

16. Memo for the record—Triviz, 1950, Box 499/Comments, Anderson.

17. T. R. Gillenwaters to IPK/File, November 20, 1950, Box 499/Comments; Triviz memo, November 21, 1950, Box 522/Working Drafts; Memo for the record, December 7, 1950, Box 499/Comments, Anderson.

18. Triviz to Anderson, December 4, 1950; H. J. Res. 550, 81st Congress, 2nd Session, December 6, 1950, Box 498/Weather Control Inquiries, Anderson.

19. Congressional Record—Senate, Introduction of Weather Control Act of 1951, December 8, 1950, p. 16475.

20. G. Francis Nauheimer to Anderson, December 11, 1950, Box 498/Inquiries, Anderson.

21. Edward G. Spottswood to Anderson, December 13, 1950, Box 498/Inquiries, Anderson.

22. Lewis C. Smith, Jr., to Anderson, December 20, 1950, Box 498/Inquiries, Anderson.

23. John Kellogg to Anderson, 1950, Box 498/Inquiries, Anderson.

24. Landsberg to Charles F. Brown, December 26, 1950, Box 480/Legislation 1950–51, RDB; H. G. Houghton to Anderson, December 29, 1950; Bernhard Haurwitz to Anderson, January 4, 1951, Box 499/Comments, Anderson; Brown to Richard A. Buddeke, undated, Box 580/Legislation 1950–51, RDB.

25. C. E. Buell to Anderson, January 5, 1951, Box 499/Comments, Anderson.

26. S. 222, 82nd Congress, 1st Session, January 8, 1951; Maurice Spears to Anderson, Box 520/General Correspondence, Anderson.

27. National Weather Improvement Association General News Release, "'Cloud Seeding' Groups Hold Convention—NWIA," January 17, 1951, Box 520/S. 222 Correspondence-NWIA, Anderson.

28. Jim Wilson to Anderson, Box 520/S. 222 Correspondence-NWIA, Anderson.

29. Anderson to Robert McKinney, January 18, 1951, Box 520/General Correspondence, Anderson.

30. Anderson to Albert Mitchell, January 18, 1951, Box 520/General, Anderson.

31. "Clint Anderson Blasts Greed in Cloud-Seeding Efforts; Pushes Own Measure for Federal Control," *Santa Fe New Mexican* 102 (51), January 24, 1951, p. 1, Box 521/Weather Control

Materials for *Rotarian* article, Anderson. The New Mexico School of Mines was renamed the New Mexico Institute of Mining and Technology in 1951.

32. "The Rain Makers," *Santa Fe New Mexican* editorial, January 26, 1951, Box 521/General Publicity, Anderson.

33. "Anderson Warns of Dangers in Seeding Clouds," January 24, 1951, Box 521/General Publicity, Anderson.

34. "Cloud Seeding Shares Blame for NM Drought," January 25, 1951, Box 521/General Publicity, Anderson.

35. "Krick Slams Clint, EDC on Rain Deal," January 25, 1951, Box 520/General Correspondence, Anderson.

36. Gillenwaters to Anderson, February 6, 1951, Box 520/S. 222 Rainmaking Correspondence from the Water Resources Development Corporation [WRDC], Anderson.

37. Anderson to Gillenwaters, February 8, 1951, Box 520/S. 222 Rainmaking Correspondence from WRDC, Anderson.

38. "Cloud Seeding Safe, Won't Bring Drought, Krick Tells Woolmen," *Santa Fe New Mexican*, February 8, 1951, Section B, Box 521/General Correspondence, Anderson.

39. "Clint Blasts State Stand on Rain Men," *Santa Fe New Mexican*, March 1, 1951, 1; "Wait for the Facts," *Santa Fe New Mexican* editorial, March 1, 1951, Box 521/General Publicity, Anderson.

40. New Mexico EDC Remarks on Senate Bill 219, Charitable Rainmaking Bill by Mc-Kinney, March 3, 1951, Box 520/S. 222 Rainmaking—Correspondence from EDC, Anderson; McKinney to Anderson, March 3, 1951, Box 521/S. 222 Rainmaking Correspondence from EDC, Anderson; "Non-Profit Rainmakers Win 'Do Pass' Senate Nod," *Santa Fe New Mexican*, March 4, 1951, 1.

41. McKinney to Anderson, March 9, 1951, Box 520/S. 222 Rainmaking Correspondence from EDC, Anderson. The possibility of having research funds held hostage by politicians in exchange for silence during hearings of scientific importance was one reason scientists had been wary of being dependent on the state (in this case, not the nation-state, but New Mexico) for research monies.

42. Maria Chabot to Anderson, undated, Box 520/General Correspondence, Anderson.

43. Gerald B. Greeman to Anderson, January 25, 1951, Box 520/General Correspondence, Anderson.

44. Resolution adopted at the Stockman and Farmers meeting, January 27, 1951, Box 520/General Correspondence, Anderson.

45. Hugh B. Woodward to Anderson, January 24, 1951, Box 520/General Correspondence, Anderson.

46. Costilla County Farm Bureau, Inc., to Anderson, January 31, 1951, Box 520/General Correspondence, Anderson.

47. C. W. Morgan, "The Rain-Making Issue," editorial, *Alamogordo (New Mexico) News,* January 1951, Box 520/General Correspondence, Anderson.

48. "Warns Bungling of Rainmaking Holds Disaster," circa January 1951, Box 521/General Correspondence, Anderson.

49. Amendment to H.R. 6 and S. 5, January 1951; S. 798, 82nd Congress, 1st Session, February 5, 1951.

50. E. C. Johnson to Joseph C. O'Mahoney, February 6, 1951, Box 520/General Correspondence, Anderson.

51. Francis Case to Anderson et al., February 26, 1951, Box 512/S. 798; Suits to Anderson, February 23, 1951; R. W. Larson to Anderson, February 27, 1951; William W. Jenkins to Anderson, March 5, 1951, Box 520/S. 222 Correspondence with GE, Anderson.

52. Anderson to Earl Platt, February 26, 1951, Box 520/S. 222 Rainmaking Correspondence from WRDC; Platt to Anderson, March 19, 1951, Box 520/General Correspondence; Anderson to Richard Searles, February 26, 1951, Box 521/General Correspondence, Anderson.

53. Buddeke to Sen. Edwin C. Johnson, February 21, 1951, Box 459/8, RDB.

54. Statement of Senator Clinton P. Anderson, March 14, 1951, p. 11, "Weather Control and Augmented Potable Water Supply," Hearings of the Subcommittee on S. 222, Committee on Interstate and Foreign Commerce, Senate; Subcommittee on S. 5, Committee on Interior and Insular Affairs, Senate; Subcommittee on S. 798, Committee on Agriculture and Forestry, Senate [hereafter Weather Control-Joint Hearings].

55. Statement and Testimony of W. F. McDonald; Reichelderfer to Joint Subcommittee, April 7, 1951, Weather Control-Joint Hearings, March 14, 1951, 12–28.

56. Testimony of C. G. Suits, 48–56; Testimony of Vincent Schaefer, 56–61; Testimony of Bernard Vonnegut, 61–69, Weather Control-Joint Hearings, March 14, 1951.

57. Testimony of Oscar L. Chapman, 75–83; Testimony of Richard D. Searles, 69–74, Weather Control-Joint Hearings, March 15, 1951.

58. Testimony of William E. Warne, Weather Control-Joint Hearings, March 15, 1951, 83–92.

59. Testimony of N. B. Bennett, Jr., Weather Control-Joint Hearings, March 15, 1951, 102–10.

60. Weather Control-Joint Hearings, March 14, 1951, 64–66.

61. Statement and Testimony of Robert McKinney, 117–22; Testimony of Carlton P. Barnes, 123–25; Testimony of Arthur A. Brown, 126–30; Testimony of Wallace E. Howell, 132–48, Weather Control-Joint Hearings, March 15–16, 1951.

62. Testimony of Vannevar Bush, Weather Control-Joint Hearings, March 16, 1951, 148–51.

63. Statement of Henry G. Houghton, Weather Control-Joint Hearings, March 16, 1951, 151–65.

64. Lee A. DuBridge, interview by Judith R. Goodstein, Pasadena, California, February 19, 1981, Part 1, 5–6, Oral History Project, California Institute of Technology Archives. Created from revised version published in *Physics in Perspective* 5 (2003): 174–205, http://resolver.caltech.edu/CaltechOH:OH_DuBridge_1.

65. Statement and Testimony of Irving P. Krick, Weather Control-Joint Hearings, April 5, 1951, 276–310.

66. "Silver-Iodide Rain Test called Threat to Defense," *NYT*, March 16, 1951, 25.

67. Brigadier General Robert E. L. Eaton to Department of the Army, Chief of Legislative Liaison, March 15, 1951; Rear Admiral T. A. Solberg to Judge Advocate General, March 26, 1951; Col. J. W. Huysoon to Rear Admiral H. A. Houser, April 20, 1951; Memo for Mr. John C. Adams, Mr. Larkin, and Mr. Leva, April 27, 1951, Box 459/8, RDB.

68. Memo for Major General Miles Reber, Rear Admiral George L. Russell, and Brigadier General Robert E. L. Eaton, probably from Houser, circa April 1951, Box 459/8, RDB.

69. Eric A. Walker to Members, RDB, May 25, 1951, Box 459/8, RDB.

70. Landsberg memo for the record, June 15, 1951, Box 459/8, RDB.

71. Ibid.; Raymond J. Burke to Morrissey, June 21, 1951; James J. Thornton to Morrissey, June 21, 1951, Box 459/8, RDB.

72. GE official to Morrissey, June 5, 1951; Morrissey to William Jenkins, July 26, 1951, Box 459/8, RDB.

73. Jim Wilson to Sen. Magnuson, June 25, 1951, Box 520/General Correspondence, Anderson.

74. H.E. 1180, 82nd Congress, 1st Session, Calendar No. 890; "Rainmaking Inquiry Sought," *NYT*, October 9, 1951, 42.

75. S. 2225, 82nd Congress, 1st Session, October 8, 1951, Box 520/S. 2225 Rainmaking Bill by Senator Case and others, Anderson.

76. Wadsworth Likely, "Bill to Permit Rain-making for 2 Years Offered," *Albuquerque Tribune*, October 10, 1951, Box 520/S. 2225, Anderson.

77. Case to Anderson, July 20, 1951, Box 520/General Publicity, Anderson.

78. Francis Schmidt to Rep. Dempsey, May 19, 1951, Box 520/General Correspondence, Anderson.

79. L. J. R. DeVries to Senator Case, July 17, 1951, Box 520/General Publicity, Anderson.

80. J. C. Shenot to Anderson, January 1952, Box 520/S. 2225, Anderson.

81. Walter Goodhue to Anderson, January 18, 1952, Box 520/S. 2225, Anderson.

82. Floyd Root to Anderson, January 29, 1952, Box 520/S. 2225, Anderson.

83. Case to Floyd Root, February 7, 1952, Box 520/S. 2225, Anderson.

84. Frank Pace, Jr., to Rep. Robert Crosser, August 7, 1951; Pace to Hon. Frederick J. Lawton, August 8, 1951, Box 459/8, RDB; Pace to Hon. Edwin C. Johnson, March 13, 1952, Box 520/S. 2225, Anderson; Walter G. Whitman memo for Assistant Secretary of Defense for Legal and Legislative Affairs, circa March 1952; Captain E. C. Stephan to Major General Miles Reber, Box 459/7, RDB.

85. H.R. 7325, 82nd Congress, 2nd Session, March 31, 1952; H.R. 7785, 82nd Congress, 2nd Session, May 8, 1952; "Rainmaking Inquiry Bill Gains," *NYT*, July 1, 1952, 25; Press release from the office of Sen. Magnuson, May 2, 1952, Box 146/172 Magnuson; Senate 82nd Congress, 2nd Session, Report No. 1514, May 12, 1952.

86. Report to Accompany S. 2225 Creating a Committee to Study and Evaluate Public and Private Experiment in Weather Modification, June 30, 1952, Box 520/S. 2225, Anderson; Senate 82nd Congress, 2nd Session, Report No. 1514, May 12, 1952.

87. Excerpt from the Congressional Record, June 2, 1952, Box 520/S. 2225, Anderson.

88. Frank J. Sherlock to Counsel, RDB, September 8, 1952, Box 459/7, RDB.

89. Harold L. Pearson to Secretary of Defense, October 8, 1952; R. L. Gilpatric to Assistant Secretary of Defense, October 8, 1952; E. C. Stephan to Assistant Secretary of Defense, October 9, 1952; Buddeke to Charles Tannenbaum, October 15, 1952; Memo for Mr. Adams on DoD 83–69, September 24, 1952, Box 4597, RDB.

90. H. R. 1064, 83rd Congress, 1st Session, January 6, 1953; S. 285, 83rd Congress, 1st Session, January 9, 1953, Senators Smathers, Magnuson, Lehman, and Butler co-sponsors.

91. W. L. Jones to Tannenbaum, February 25, 1953; H.R. 1584, 83rd Congress, 1st Session, January 13, 1953; Frank J. Sherlock to Counsel, RDB (and others), February 24, 1953; H.R. 2580, 83rd Congress, 1st Session, February 3, 1953; Sherlock to Counsel, RDB (and others),

March 11, 1953; S. 285 Union Calendar No. 378, 83rd Congress, 1st Session, June 8, 1953; Public Law 256 83rd Congress, Chapter 425, 1st Session, S. 285, August 13, 1953, Box 459/7 RDB.

92. "U.S. to Do Something about the Weather," *NYT*, December 10, 1953, 1.

93. Charles E. Egan, "Weather Visioned as Weapon of U.S.," *NYT*, December 11, 1953, 49.

94. Neal Hanson to Anderson, March 12, 1954; Cash Ramey to Charles Gardner, Jr., March 12, 1954; Pedro Medina to Anderson, May 25, 1954, Box 545/S. 285; "Ban on Rain-Making Efforts Asked by Group in Raton," *Albuquerque Journal*, July 17, 1954, Box 498/ Background Materials Weather Control Act; E. T. Springer to Anderson, July 17, 1954, Box 520/General Correspondence; Anderson to Neal Hanson, March 17, 1954, Box 545/S. 285, Anderson.

95. Karl T. Pfeister to Anderson, and J. L. Hinman to Anderson, ca. spring 1954, Box 520/ General Correspondence, Anderson.

96. Congressional Record—U.S. Senate, July 22, 1954, 10917.

97. Sen. Francis Case, "Does 'Rain-Making' Pay? National Body Set Up to Check on 'Cloud Modification,'" June 1954, Box 2381/Weather Control Advisory Committee Apr-Aug 1954, SecAg.

98. Frank Carey, "Congress Keeps an Eye on Rainmaking," Washington *Evening Star*, July 2, 1954, Box 498/Background Materials Weather Control Act, Anderson.

99. Using insecticides developed during World War II, the USDA tried to control imperfect nature by tackling fire ants to improve the agriculture sector. That effort also eliminated many other, desirable creatures as an unexpected consequence of a technological fix. See Joshua Blu Buhs, *The Fire Ant Wars: Nature, Science and Public Policy in Twentieth-Century America* (Chicago: University of Chicago Press, 2004). On insect eradication, see John H. Perkins, "Insects, Food, and Hunger: the Paradox of Plenty for U.S. Entomology, 1920–1970," *Environmental Review: ER* 7, no. 1 (1983): 71–96. On herbicides, which had also gotten a research boost during World War II, and their use in agriculture, see J. L. Anderson, "War on Weeds: Iowa Farmers and Growth-Regulator Herbicides," *Technology and Culture* 46, no. 4 (2005): 719–44.

100. True D. Morse, April 2, 1954, Box 2381/Weather Control Advisory Committee Apr-Aug 1954, SecAg. On postwar control of national forests, see Paul W. Hirt, *A Conspiracy of Optimism: Management of the National Forests since World War II* (Lincoln: University of Nebraska Press, 1994). See also Christopher McGrory Klyza, *Who Controls Public Lands? Mining, Forestry, and Grazing Policies, 1870–1990* (Chapel Hill: University of North Carolina Press, 1996). On fire suppression, see Stephen J. Pyne, *Fire in America: A Cultural History of Wildland and Rural Fire* (Seattle: University of Washington Press 1997). Pyne discusses the conversion of military hardware/research in fire physics and fire weather to civilian use after World War II as part of the Civil Defense program.

101. Leonard A. Scheele to Orville, April 8, 1954, Box 2381/Weather Control Advisory Committee Apr-Aug 1954, SecAg. On the control of mosquitoes and malaria in the US South, see Margaret Humphreys, "Kicking a Dying Dog: DDT and the Demise of Malaria in the American South, 1942–1950," *Isis* 87, no. 1 (1996): 1–17; Margaret Humphreys, *Malaria, Poverty, Race, and Public Health in the United States* (Baltimore: Johns Hopkins University Press, 2001); Pete Daniel, *Toxic Drift: Pesticides and Health in the Post-World War II South* (Baton Rouge: Louisiana University Press in cooperation with the Smithsonian National Museum of Amer-

ican History, 2005); and Thomas R. Dunlap, "Science as a Guide in Regulating Technology: The Case of DDT in the United States," *Social Studies of Science* 8, no. 3 (1978): 265–85. On keeping insects of all kinds under control with chemicals that had been developed by military services starting with World War I, see Edmund Russell, *War and Nature: Fighting Humans and Insects with Chemicals from World War I to Silent Spring* (Cambridge: Cambridge University Press, 2001).

102. Director, USGS, to Under Secretary (Interior), May 4, 1954, Box 2381/Weather Control Advisory Committee Apr-Aug 1954, SecAg.

103. H. F. McPhail to Ralph A. Tudor, May 4, 1954; John L. Farley to Secretary of the Interior, May 3, 1954; Edward Woozley to Tudor; and Morgan D. Dubrow to Tudor, May 5, 1954, Box 2381/Weather Control Advisory Committee Apr-Aug 1954, SecAg.

104. Draft report of the President's ACWC concerning DoD activities in the field, April 12, 1954, Box 2381, SecAg.

105. True D. Morse, April 2, 1954; H. F. McPhail to Ralph A. Tudor, May 4, 1954; Leonard A. Scheele to Orville, April 8, 1954; John L. Farley to Interior Secretary, May 3, 1954; Edward Woozley to Tudor; and Morgan D. Dubrow to Tudor, May 5, 1954, all from Box 2381, SecAg.

106. Robert B. Murray, Jr., to Orville, July 2, 1954, Box 2381/Weather Control Advisory Committee Apr-Aug 1954, SecAg.

107. Thomas H. Miller to Tudor, May 3, 1954, Box 2381/Weather Control Advisory Committee Apr-Aug 1954, SecAg.

108. National Park Service Interest in Artificial Weather Control, May 1954, Box 2381/Weather Control Advisory Committee Apr-Aug 1954, SecAg. On conservation and preservation in the national parks, particularly when attempts to control nature would impinge on park land, see Neil M. Maher, *Nature's New Deal* (Oxford: Oxford University Press, 2008), 220–24. Also see Mark Harvey, *A Symbol of Wilderness: Echo Park and the American Conservation Movement,* 2nd edition (Seattle: University of Washington Press, 2000); and Roderick Nash, *Wilderness and the American Mind* (New Haven, CT: Yale University Press, 1967), 209–19.

109. Statement of Interest of the NSF in Weather Control, April 1954, Box 2381/Weather Control Advisory Committee Apr-Aug 1954, SecAg.

CHAPTER FOUR

1. "Amateur Rain Makers," from *Indianapolis Star* in the *Christian Science Monitor,* December 2, 1950, 18, Box 498/Background Material Weather Control, Anderson.

2. "Expert to Survey Rain-Making Plan," *NYT*, February 21, 1950, 9.

3. "The Legality of Rainmaking," *NYT*, March 2, 1950, 26.

4. Charles G. Bennett, "City Speeds Plan for Rain Making," *NYT*, March 1, 1950, 37.

5. Charles G. Bennett, "Success of Dry Day Defers Curbs; Mayor Picks Rain-Making Advisers," *NYT*, March 4, 1950, 1; Francis Reichelderfer to Victor F. Hess, March 13, 1950, Box 459/8, RDB.

6. Charles G. Bennett, "Rain-making Tests Begin within Week; $50,000 Fund Voted," *NYT*, March 15, 1950, 1.

7. "$50,000 to Make Rain," *NYT*, March 15, 1950, 28.

8. Francis W. Reichelderfer, "Letters to the Times: Methods to Induce Rain," *NYT*, March 27, 1950, 22.

9. Charles G. Bennett, "Rain-Maker Seeks Mountain 'Office,'" *NYT*, March 16, 1950, 33.

10. "Radar Will Check City's Rain Tests," *NYT*, March 22, 1950, 3.

11. Charles G. Bennett, "Suit Seeks to Block Rain-Making but City Will Not Halt its Tests," *NYT*, March 23, 1950, 1.

12. "Rain-Praying Preacher asks $7000 from City," *NYT*, March 28, 1950, 33.

13. Paul Crowell, "Supply of Dry Ice Ready for Clouds," *NYT*, March 26, 1950, 1; "Storage of Water Inches up to 55.4%," *NYT*, March 27, 1950, 1; "Rain Men Stayed by Fear of Flood," *NYT*, April 2, 1950, 6.

14. Charles G. Bennett, "Water Supply up 7 Billion Gallons," *NYT*, March 30, 1950, 31.

15. "World News Summarized," *NYT*, April 14, 1950, 1.

16. Charles G. Bennett, "'Howell's Snow' Irks Some but City Calls It Fine Stuff," *NYT*, April 15, 1950, 1.

17. "World News Summarized," *NYT*, April 21, 1950, 1.

18. "Rainmaking Tests Seen Inconclusive," *NYT*, April 27, 1950, 31.

19. "Crew Tries Again to Smoke Out Rain," *NYT*, May 2, 1950, 28.

20. "City Allays Fears on Rain-Making: Effects, If Any, Only in Catskills," *NYT*, May 18, 1950, 31.

21. "Water Gains Shown on Every Front," *NYT*, May 20, 1950, 17; "Dry Day Failure, but Supplies Gain," *NYT*, May 27, 1950, 13.

22. "Heavy Rains Fill Catskill System," *NYT*, June 5, 1950, 23.

23. "Catskill System Still Overflowing," *NYT*, June 6, 1950, 31. Dutchess County is north of the city.

24. "Water Supply Off as City Seeks Rain," *NYT*, July 11, 1950, 33; "Car-Washing Ban May End This Week," *NYT*, July 11, 1950, 33.

25. "Rainmaking Suit Threatened," *NYT*, July 11, 1950, 33.

26. "More Rain-Making Backed by Carney," *NYT*, July 21, 1950, 38.

27. "Rain-maker Gets Another 6 Months," *NYT*, August 18, 1950, 21.

28. "Rainmaking Splits State's Farmers," *NYT*, August 23, 1950, 31.

29. "Rainmaking Curbs Asked," *NYT*, January 24, 1951, 22; "The Proceedings in Albany," *NYT*, January 30, 1951, 18.

30. "City Sued in Rain-Making," *NYT*, February 17, 1951, 8.

31. "Upstate Sues NYC for $288,333 in 'Rain-Making,'" *Christian Science Monitor*, February 28, 1951, 14; "Kingston Group Sues for Damage by Rains," *NYT*, June 30, 1951, 13.

32. "City Now Skeptic on Rain-Making; Damage Claims Total $2,138,510," *NYT*, November 5, 1951, 1.

33. "Suits Pour on City for Its Rainmaking," *NYT*, December 22, 1951, 1.

34. "Rainmaking Abandoned," *NYT*, November 10, 1951, 16.

35. "City Won't Resort to Artificial Rain," *NYT*, April 28, 1953, 18.

36. "The Proceedings in Albany," *NYT*, March 14, 1953, 9.

37. "US Will Try Rainmaking in Power Crisis," *Washington Post*, September 26, 1951, 1. See also Thomas E. Mullaney, "Northwest Will Conserve Power to Keep Aluminum Plants There," *NYT*, September 30, 1951, 103.

38. Phil Church to Lloyd S. Woodburne, October 5, 1951, Box 2/Cloud Seeding, Church.

39. Raymond B. Allen to Fred Koch, October 17, 1951, Box 2/Restricting Weather Modification, Church.

40. Transcript of Hearing by the Subcommittee on Natural Resources on Artificial Cloud Stimulation, December 15, 1951, Box 2/Restricting, Church. On Frederick Newell and irrigation, see Donald J. Pisani, *Water and American Government: The Reclamation Bureau, National Water Policy, and the West, 1902–1935* (Berkeley: University of California Press, 2002), 2.

41. Transcript of Hearing by the Subcommittee on Natural Resources on Artificial Cloud Stimulation, December 15, 1951, Box 2/Restricting, Church.

42. Ibid.

43. Ibid.

44. Rep. Hodde, Transcript of Hearing by the Subcommittee on Natural Resources on Artificial Cloud Stimulation, December 15, 1951, Box 2/Restricting, Church.

45. Robert Bernethy to Phil Church, May 15, 1952; Church to Bernethy, May 19, 1952, Box 2/Cloud Seeding, Church.

46. Transcript of Hearing by the Subcommittee on Natural Resources on Artificial Cloud Stimulation, June 7, 1952, 1–2, Box 2/Restricting, Church.

47. Ibid., 4, 7.

48. Ibid., 6.

49. Ibid., 8.

50. Ibid., 11–12, 18.

51. Wilbur G. Hallauer to Church, June 26, 1952, Box 2/Restricting, Church.

52. Church to Hallauer, July 1, 1952, Box 2/Restricting, Church.

53. "Rain War Over, but Who Won," *Seattle Daily Times*, July 22, 1952, Box 2/Cloud Seeding, Church.

54. Church to Gov. Arthur B. Langlie, November 18, 1952, Box 2/Restricting, Church.

55. "Mountain States: Semi-Arid Regions Look with Hope to Rain Experiments," *NYT*, April 2, 1950, 146.

56. "The Southwest: All Aspects of Rainmaking Are Taken Up for Study," *NYT*, June 4, 1950, E8.

57. "Snow and Rain Laid to Seeding of Clouds," *NYT*, October 14, 1951, 86.

58. Act Number 53, Senate, 31st Legislature (Wyoming), February 19, 1951; passed from Mr. Gardner to Senator Case, to Mr. Morrissey, June 12, 1951, Box 459/8, RDB.

59. "Wyoming Scores 'First' in Cloud Seeding Permits," *Christian Science Monitor*, May 21, 1951, Box 521/General Publicity, Anderson.

60. W. McNab Miller and Verne J. Varineau, "Verification Studies of Commercial Cloud Seeding Operations in Wyoming for the Period 1 May 1951 through 30 September 1951 (Progress Report)," Wyoming Experiment Station—Mimeograph Circular No. 8, November 1951.

61. "Rain-Making Is Assailed," *NYT*, June 13, 1951, 30.

62. Ted Driscoll to John C. Morrissey, February 11, 1952, Box 459/7, RDB.

63. State of California Senate Bill No. 617, July 23, 1951, Box 2/Restricting, Church.

64. Oregon Legislative Interim Committee on Weather Control, November 22, 1952, Box 2/Restricting, Church; State of Oregon, "Weather Modification," Report of Interim Committee

Appointed Pursuant to the House Joint Resolution No 22, 46th Legislative Assembly, Box 45/ State of Oregon, E/ARM.

65. Oregon Senate Bill No. 103, Box 2/House Bills 279, 195; R. T. Beaumont to Oregon State Senator Ben Day, February 3, 1953, Box 2/Restricting; Oregon House Bill No. 266, February 4, 1953, Box 2/House Bills 279, 195, Church.

66. R. T. Beaumont, "An Appraisal of Cloud-Seeding and Recommendations for Further Study," 1953, Box 2/Restricting, Church.

67. Driscoll to Morrissey, February 11, 1952, Box 459/7, RDB.

68. Stuart A. Cundiff to Langmuir, Box 6/1951, Langmuir.

69. "In New Mexico," *Albuquerque Journal*, May 11, 1951, Box 532/General Publicity, Anderson.

70. Eddie Lee, "Rain-Makers Gain Hopefuls," *Santa Fe New Mexican,* May 16, 1951, Box 521/General Publicity, Anderson.

71. "Seeding Caused Big Rain—Krick," *Albuquerque Tribune,* May 17, 1951, Box 521/General Publicity, Anderson.

72. "Farmers Cry 'Uncle' on Rain-Making," *Santa Fe New Mexican,* May 25, 1951, Box 521/General Publicity, Anderson.

73. "Dr. Krick Declares NM Drought Broken," *Hobbs Daily News Sunday*, July 27, 1951, Box 520/General Publicity, Anderson.

74. Anderson to Krick, May 22, 1951, Box 520/S. 222 Correspondence—NWIA, Anderson.

75. "Must Have Clouds First, Krick Says," *Albuquerque Journal*, August 16, 1951, Box 520/General Publicity, Anderson.

76. Robert D. Elliott to Anderson, July 27, 1951, Box 520/S. 222 Rainmaking Bill, Anderson.

77. Elliott to Langmuir, August 1, 1951, Box 6/1951, Langmuir.

78. Talk by T. H. Hazzard to the meeting of Idaho Reclamation officials, May 23, 1951, Entry 120, Box 5/Practices of Commercial Rainmakers, USWB.

79. Ibid.

80. Comments on Hazzard's talk, May 23, 1951, Entry 120, Box 5/Practices, USWB.

81. Paul A. Humphrey to Richard E. Georgi, May 26, 1951, Entry 120, Box 5/Practices, USWB.

82. Reichelderfer to William Weber, July 16, 1951, Entry 120, Box 5/Practices, USWB.

83. Weber to Humphrey, August 16, 1951, Entry 120, Box 5/Practices, USWB.

84. Humphrey to Weber, August 23, 1951; Weber to Reichelderfer, August 27, 1951; Humphrey to Harry Wexler, September 9, 1951, Entry 120, Box 5/Practices, USWB.

85. "Cloud-icing Row Marks Rainmakers' Meeting," January 26, 1952, Box 521/General Publicity, Anderson.

86. Robert McKinney to Schaefer, October 27, 1953, Schaefer.

87. "Rainmaking Study Urged," *NYT*, July 29, 1953, 25.

88. "In the Offing," *Washington Daily News*, July 31, 1954, Box 2380/Weather Control Advisory Committee 1954, SecAg.

89. "Drought Aid Widened," *NYT*, September 30, 1954, 17.

90. Substitute House Bill No. 379, State of Washington, 33rd Regular Session, February 11, 1953; Recommendations of the Cherry Institute Board of Directors re House Bill 379, February 16, 1953; Robert Elliott to Phil Church, February 16, 1953; Church to Elliott, February 20,

1953; Church to Wilbur Hallauer, March 6, 1953; Hallauer to Church, March 11, 1953, Box 2/ House Bills 379, 1995, Church.

91. Washington House Bill 216, 34th Regular Session, January 24, 1955; Hallauer to Church, February 26, 1955; Church to Hallauer, February 24, 1955, Box 2/House Bills 375, 195, Church.

92. House Bill 1995, State of Washington, 35th Regular Session, Box 2/House Bills 375, 195; Chapter 245, House Bill 195 (State of Washington) State Weather Modification Board, signed by the Governor of Washington, March 23, 1957, Box 2/Reports, Church.

CHAPTER FIVE

1. Henry Houghton, Comments on Weather Modification, circa July 1954, Box 2380/ Weather Control Advisory Committee 1954, SecAg.

2. Harry Wexler, "Precipitation—Natural or Otherwise," September 20, 1951, Box 34/ Weather Modification, Wexler. For a similar point of view, see C. R. Holmes and William Hume II, "Final Research Report to the State of New Mexico, Economic Development Commission—Water Resources Development," New Mexico Institute of Mining and Technology (R&D Division), Socorro, New Mexico, September 1951.

3. The meteorologists' attempts to influence the discussion, particularly through the American Meteorological Society, fit the pattern described by Brian Balogh as being part of the "Associational State."

4. AMS News Release, March 16, 1951, Box 459/8, RDB.

5. Henry Harrison to Henry G. Houghton, copy to Sverre Petterssen, February 3, 1951, Box 1/AMS Cloud Seeding Committee, 1951, Houghton. On the disciplinary status of meteorology, see Kristine C. Harper, *Weather by the Numbers: The Genesis of Modern Meteorology* (Cambridge, MA: MIT Press, 2008), chap. 3. Concerns about the effects of secrecy, particularly related to military scientific efforts, became especially acute after World War II ended and the Cold War began. See, e.g., Michael Aaron Dennis, "Secrecy and Science Revisited: From Politics to Historical Practice and Back," in *The Historiography of Contemporary Science, Technology, and Medicine: Writing Recent Science,* ed. Ronald E. Doel and Thomas Söderqvist (London, 2006), 172–84; and Jessica Wang, *American Science in the Age of Anxiety: Scientists, Anticommunism, and the Cold War* (Chapel Hill: University of North Carolina Press, 1999).

6. Harrison to Houghton, February 3, 1951, Box 1/AMS 1951, Houghton.

7. Harrison to Houghton, March 19, 1951, Box 1/AMS 1951, Houghton. The bills in question were S. 5, S. 222, and S. 798, 82nd Congress, 1st Session.

8. Houghton to Harrison, March 23, 1951, Box 1/AMS 1951, Houghton.

9. Houghton to Petterssen and Harrison, with enclosure, March 29, 1951, Box 1/AMS 1951, Houghton.

10. Wadsworth Likely, "Plan for Umpires in Rain Making Dispute Reported," Science Service article, May 3, 1951, Box 521/General Publicity, Anderson.

11. Notes for United Airlines Meteorologists, "Further Developments in Weather Control," by H. T. Harrison, Manager of Weather Services; Harrison to Houghton, July 3, 1951, Box 1/ AMS 1951, Houghton.

12. Houghton to Petterssen and Harrison, July 12, 1951, Box 1/AMS 1951, Houghton.

13. Harrison, Petterssen, and Houghton to Charles F. Brooks, August 15, 1951, Box 1/AMS 1951, Houghton.

14. AMS Statement on the Legislative Aspects of Weather Modification, October 1951, Box 2/Cloud Seeding, Church.

15. Harrison to Houghton, February 12, 1952, Box 1/AMS 1951–1953, Houghton.

16. Harrison to Houghton, July 8, 1952, Box 1/AMS 1951–1953, Houghton.

17. Petterssen to Houghton, September 30, 1952, Box 1/AMS 1951–1953, Houghton.

18. Harrison to Houghton, October 28, 1952, Box 1/AMS 1951–1953, Houghton.

19. Brooks to Houghton, March 5, 1953, Box 1/AMS 1951–1953, Houghton.

20. Harrison, Petterssen, and Houghton to Brooks, March 10, 1953, Box 1/AMS 1951–1953, Houghton.

21. Eugene Bollay and Robert D. Elliott to AMS Council, May 14, 1953, Box 1/AMS 1951–1953, Houghton. They were referring to Manes Barton, "Analysis of Attempts to Increase Snow-fall and Resultant Runoff by Winter Cloud Seeding in Southern Oregon Cascade Mountains, 1951–52," Cultural Research Foundation, Corvallis, Oregon, ca. 1952. Harrison to Houghton, December 3, 1953, Box 1/AMS 1951–1953, Houghton.

22. W. O. Senter to Brooks, June 16, 1953, Box 1/AMS 1951–1953, Houghton.

23. Houghton to Petterssen and Harrison, August 13, 1953, Box 1/AMS 1951–1953, Houghton.

24. Here we see the scientists playing the "USSR is ahead of us" card to get more funding. See, e.g., David J. Tietge, *Flash Effect: Science and the Rhetorical Origins of Cold War America* (Athens: Ohio University Press, 2002). For additional sources, see the notes associated with Part II's introduction.

25. Petterssen to RADM C. M Bolster, November 9, 1951 [SECRET], Box 459/8, RDB. In addition to Petterssen, other committee members included Erwin R. Biel, Charles L. Critch-field, Daniel F. Rex, H. P. Robertson, H. J. Stewart, Alan T. Waterman, and Max A. Woodbury.

26. Ibid.

27. A. I. Voskresenskiy, V. G. Morachevskiy, V. Ya. Nikandrov, "The Use of Dry Ice for Dispersing Clouds in the Arctic," *Problems in the Arctic, A Collection of Articles* 2 (1957): 133–39; Yu. A. Seregin, "Dispersion of Supercooled Fogs from the Ground with Silver Iodide Aerosol," *Trudy Tsentral. Aerolog. Observatorii* 19 (1958): 68–80. See also V. J. Nikandrov, "The Problem of Artificial Control of Cloud and Fog" (1958), and I. I. Gaivoronskii, "Investigation Conducted at the Central Aerological Observatory on Artificial Modification of Clouds and Fog" (1959). Gaivoronskii wrote that controlling weather was "one of the most important tasks of contemporary science." A. P. Chuvaev, "O sovremennykh vozmozhnostyakh predotvrashcheniya groz i grada (Modern Possibilities of Preventing Storm and Hail)," *Trudy GGO* 74 (1957).

28. On Russian efforts in the Arctic, see Paul R. Josephson, *The Conquest of the Russian Arctic* (Cambridge, MA: Harvard University Press, 2014).

29. Moscow UPI dispatch, November 25, 1957, Box 35/Weather Modification, Rosellini.

30. UPI Dispatch, December 23, 1957, Box 35/Weather Modification, Rosellini.

31. Hanson W. Baldwin, "Major US Effort Urged to Match Soviet Strength," *NYT*, February 5, 1958, 1.

32. Public Law 256, 83rd Congress, Chapter 426, 1st Session, approved August 13, 1953.

33. Charles F. Willis, Jr., to Honorable Leonard Hall (RNC); Willis to Charles E. Wilson;

Willis to Douglas McKay; Willis to Ezra Taft Benson, all of September 4, 1953; Willis to Oveta Culp Hobby; Willis to Alan Waterman, both of September 14, 1953, Box 746/OF 244 Advisory Committee on Weather Control (ACWC) (1), DDE/WHCF.

34. Wilson to Willis, September 22, 1953; Ezra Taft Benson to Willis, October 7, 1953; Senator Francis Case to Willis, October 16, 1953, Box 746/OF 244-1, DDE/WHCF.

35. For example, Willis to Sinclair Weeks, October 19, 1953, Box 746/OF 244-1, DDE/WHCF.

36. Willis to Governor Sherman Adams, December 2, 1953, Box 746/OF 244-1, DDE/WHCF.

37. Orville to President Eisenhower, Enclosure B, June 10, 1954, Box 746/OF 244-1, DDE/WHCF.

38. Howard T. Orville, Statement on Senate Bill S. 86, March 26, 1957; Interim Report of the Advisory Committee on Weather Control, February 8, 1956, Orville.

39. Carleton P. Barnes to J. Earl Coke and Byron T. Shaw, February 15, 1954; ACWC Meeting, February 13, 1954; ACWC Meeting Minutes, February 29, 1954; Orville to ACWC Members, March 12, 1954: J. Vern Hales, Thomas E. Hoffer, and Eugene L. Peck, "Evaluation of Rain Making Experiment in Utah," presented at the Western Snow Conference, Salt Lake City, Utah, April 19, 1954, Box 2381/Weather Control Advisory Committee Jan-Mar 1954, SecAg.

40. Luna B. Leopold to Chairman, ACWC, May 21, 1954, Box 2381/Weather Control Jan-Mar 1954, SecAg.

41. Statement of Interest of the National Science Foundation in Weather Control, April 1954, Box 2381/Weather Control Jan-Mar 1954, SecAg.

42. Minutes of the 4th Meeting—ACWC, May 4, 1954, Box 2381/Weather Control Apr-Aug 1954, SecAg.

43. Orville to President Eisenhower, letter with attachments, June 10, 1954, Box 746/OF 244-2, DDE/WHCF.

44. ACWC Minutes of the 5th Meeting, August 9, 1954, Box 2380/ACWC 1954, SecAg.

45. R. R. Reynolds, N. D. Sturm, and E. P. Warren, "Cloud Seeding—A Problem of National Importance," paper given September 1952 and forwarded to the ACWC by Orville, July 14, 1954, Box 2381/ACWC Apr-Aug 1954, SecAg; North American Weather Consultants, "Preliminary Statement on Drought Suppression," submitted to the ACWC, July 1954, Box 2/Cloud Seeding, Church; Archie M. Kahan, "Review of Reports of Cloud Seeding," August 4, 1954, Box 2381/ACWC Apr-Aug 1954, SecAg; and, for example, Houghton, "Comments on Weather Modification," circa July 1954; John S. Battle to Charles Gardner, Jr.; Paul B. MacCready, Jr., "Report to the Advisory Committee on Weather Control," August 27, 1954; Richard D. Coons to Orville, August 30, 1954; Irving Krick to Gardner, August 30, 1954; Harrison, "An Informal Survey on the Status of Artificial Weather Control," circa August 1954, Box 2380/ACWC 1954, SecAg.

46. George A. Pavellis, "Agricultural Implications of Weather Control," Monograph Circular 86 (Bozeman: Montana State College, Agriculture Experiment Station, January 1955).

47. Eugene Bollay, "A Report to the Panel of the ACWC on Effects of Atomic and Thermonuclear Explosions in Modifying the Weather, November 28, 1955, Schaefer.

48. "Mount Washington Observatory News Bulletin," no. 28, December 1955, Schaefer.

49. Howard T. Orville to President Eisenhower, February 1956, Box 114/18, Magnuson.

50. First Interim Report, Advisory Committee on Weather Control, February 1956, Schaefer.

51. Reichelderfer to Waterman, February 10, 1956, Box 7/7, Reichelderfer.

52. Reichelderfer to Lewis W. Douglas, February 14, 1956, Box 7/7, Reichelderfer.

53. Gardner to Schaefer, February 28, 1956, Schaefer.

54. Extending for 2 Years the Advisory Committee on Weather Control; Report No. 1866, Calendar No. 1888, 84th Congress, 2nd Session, April 26, 1956.

55. Letter from James E. McDonald summarizing the Conference on the Scientific Basis for Weather Modification Studies, June 1, 1956, Orville.

56. H. C. S. Thom, "An Evaluation of a Series of Orographic Cloud Seeding Operations"; K. A. Brownlee, "Statistical Tests by the Method of a Control of the Rainmaking Hypothesis: Some Alternative Viewpoints"; Jerzy Neyman, "Observational Prerequisites to a Decisive Statistical Evaluation of Cloud Seeding Operations"; Max Woodbury, "The ACWC Statistical Program," all presented at the Conference on the Scientific Basis for Weather Modification Studies, Institute of Atmospheric Physics, University of Arizona, April 10–12, 1956; "Does Cloud-Seeding Produce Rain? Still Moot Question," (Tucson) *Arizona Daily Star*, April 11, 1956; John Riddick, "President's Group Defends Report on Rainmaking," *Tucson Daily Citizen*, April 13, 1956; "Rain-Making," *NYT*, April 15, 1956, Orville.

57. Letter from James E. McDonald, June 1, 1956, Orville.

58. Orville to President Eisenhower, June 28, 1957, Box 151/670-2, Anderson.

59. Percival Brundage to Sherman Adams, June 13, 1957, Box 746/OF 244-3, DDE/WHCF.

60. Telegram from Lewis Douglas to Gen. Wilton B. Persons, June 18, 1957; Douglas to Senator Carl Hayden, June 6, 1957, Box 746/OF 244-3, DDE/WHCF.

61. Orville to Jack Z. Anderson, January 23, 1958, Box 746/OF 244-9, DDE/WHCF.

62. *Final Report of the Advisory Committee on Weather Control*, vol. 1 (Washington, DC: 1958), iii.

63. Ibid., vi–vii.

64. Ibid., vii.

65. Ibid., vii–viii. On the operational aspects of numerical weather prediction in the mid-1950s, see Harper, *Weather by the Numbers*, chap. 7.

CONCLUSION TO PART TWO

1. Quoted in Dan Kevles, "Cold War and Hot Physics: Science, Security, and the American State, 1945–56," *Historical Studies in the Physical and Biological Sciences* 20, no. 2 (1990): 312.

2. Howard T. Orville, "Future Research in Weather Modification," circa 1959, Orville.

3. Jack C. Oppenheimer, Executive Secretary, ACWC, "The Legal Situation," *Final Report of the Advisory Committee on Weather Control* (Washington, DC: 1958), 213.

4. Biennial Report of the Weather Modification Board (Shumway), June 30, 1958, Box 117/Weather Modification Board 1964, E/AP/RA.

5. I. Richard Adlard, County Extension Agent, to Joe Dwyer, Director, State Department of Agriculture, June 1959, Box 55/WRDC Correspondence re Seeding Pacific Power and Light 1958–60, E/ARM.

6. Mr. and Mrs. Wesley C. Lamon to Stuart E. Shumway, September 24, 1959, Box 55/WRDC Correspondence re Seeding Pacific Power and Light 1958–60, E/ARM.

7. Mrs. Esther Dower Fleshman to Governor Rosellini, October 6, 1959, Box 1959/Weather Modification Board, Rosellini.

8. Statement to be given at hearing on cloud seeding by J. William Antilla, October 21, 1959, Box 2/Miscellany, Church.

9. "Heavy Rain Blamed for Large Losses," *Longview Daily News*, October 22, 1959, Box 1959/Weather Modification Board, Rosellini.

10. Esther Dower Fleshman to Governor Rosellini, undated circa fall 1959, Box 1959/Weather Modification Board, Rosellini.

11. Church to Walter Orr Roberts, February 1, 1960, Box 2/Outgoing Letters, Church.

12. Brian Balogh, *Chain Reaction: Expert Debate and Public Participation in American Commercial Nuclear Power, 1945–1975* (Cambridge: Cambridge University Press, 1991), 21–23.

13. Michael Adas, *Dominance by Design: Technological Imperatives and America's Civilizing Mission* (Cambridge, MA: Belknap Press, 2006), 8.

PART THREE

1. Nate Haseltine, "U.S. Seriously Concerned; Cold War May Spawn Weather Control Race," *Washington Post and Times Herald*, December 23, 1957, 1. According to Michael Adas, "Both superpowers remained deeply ambivalent about the scientific and technological capabilities of their rivals and suspicious about the uses to which these were likely to be put." Michael Adas, *Dominance by Design: Technological Imperatives and America's Civilizing Mission* (Cambridge, MA: Belknap Press, 2006), 309.

2. Haseltine, "U.S. Seriously Concerned."

3. On using science and technology as a state tool, see John Krige, *American Hegemony and the Postwar Reconstruction of Science in Europe* (Cambridge, MA: MIT Press, 2006); Walter A. McDougall, . . . *The Heavens and the Earth: A Political History of the Space Age* (Baltimore: Johns Hopkins University Press, 1997) [originally Basic Books, 1985]; Alan I. Marcus and Howard P. Segal, *Technology in America: A Brief History* (Fort Worth, TX: Harcourt Brace, 1999); Paul R. Josephson, *Resources under Regimes* (Cambridge, MA: Harvard University Press, 2004; NAS-NRC, *Weather and Climate Modification, Problems and Prospects*, vol. 1, Summary and Recommendations, Report No. 1350 (Washington, DC: NAS-NRC, 1966); Carroll Pursell, *Technology in Postwar America: A History* (New York: Columbia University Press, 2007). For some of the wilder geoengineering-type schemes, see James Rodger Fleming, *Fixing the Sky: The Checkered History of Weather and Climate Control* (New York: Columbia University Press, 2010).

4. Haseltine, "U.S. Seriously Concerned."

5. Ibid. Other articles were of this same ilk. See, e.g., Truman Temple, "Federal Group Warns of Weather Warfare, Urges Added Research; Study Suggests Russia Could Alter World's Rain, Flood U.S. by Melting Polar Ice," *Wall Street Journal*, December 11, 1957, 1; Capt. Howard T. Orville, U.S.N. (Ret.) as told to Joe Alex Morris, "Stand By for Climate Control," *Saturday Evening Post* 231, no. 22 (November 29, 1958): 19.

6. Allen W. Dulles, Memorandum for the Record, February 27, 1958 [SECRET], CREST. The CIA CREST (CIA Records Search Tool) database, which can be accessed at National Archives II, College Park, MD, contains declassified CIA documents. New nonexempt doc-

uments are added as they pass through the automatic twenty-five-year review period, under the provisions of Executive Order 13256. See http://www.foia.cia.gov/collection/crest-25-year -program-archive for additional information about this valuable resource.

7. See, e.g., Geographic Division, O/RR (Office of Research and Reports), Current Status and Plans, January 15, 1951 [SECRET]; Germany: Research on Weather Modification, 1952 [CONFIDENTIAL]; Intelligence Advisory Committee, Priority National Scientific and Technical Intelligence Objectives, December 27, 1955, and March 5, 1957 [SECRET], CREST.

8. V. A. Shtal' and V. G. Morachevskiy, "Active Influence on the Weather USSR," *Priroda* 47, no. 9 (September 1958) (translated January 1959). Historian Paul Josephson has written extensively on Soviet technoscience and the control of nature. See his *Totalitarian Science and Technology* (Amherst, NY: Humanity Books, 2005); *Resources under Regimes;* and *Industrialized Nature: Brute Force Technology and the Transformation of the Natural World* (Washington, DC: Island Press/Shearwater Books, 2002). See also David W. Keith, "Geoengineering the Climate: History and Prospect," *Annual Review of Energy and Environment* 25 (2000): 250–51.

9. This is not surprising since Bergeron had extensive contacts with Soviet meteorologists in connection with Norway's Bergen School of Meteorology. On Bergeron, see Duncan C. Blanchard, "The Life and Science of Tor Bergeron," *BAMS* 59, no. 4 (1978): 387–90.

10. E. K. Federov, "Physical Methods of Influencing the Weather," All Union Society for the Dissemination of Political and Scientific Knowledge, Moscow, 1959.

11. Ibid.

12. Ibid.

13. Austin C. Wehrwein, "Soviet Endangers U.S. Lead in Meteorology, Scientists Say," *NYT*, March 28, 1959, 6.

14. "Russian Doubtful of Weather Curb," *NYT*, July 21, 1959, 5.

15. Ch'eng Shun-shu, "China's Experiments in Artificial Precipitation," *Ch-i-hsiang Hsueh-pau (Journal of Meteorology)* XXX, no. 3 (1960): 286–90. On the relationship between science and the state in China, and how the Cold War influenced scientific efforts, see Sigrid Schmalzer, "Self-Reliant Science: The Impact of the Cold War on Science in Socialist China," in *Science and Technology in the Global Cold War,* ed. Naomi Oreskes and John Krige (Cambridge, MA: MIT Press, 2014), 75–106; Fa-ti Fan, "Science, State, and Citizens: Notes from Another Shore," *Osiris* 27 (2012): 227–49, and "Redrawing the Map: Science in Twentieth-Century China," *Isis* 98, no. 3 (2007): 524–38.

16. Chen-ch'ao Ku, "Artificial Precipitation Work in Communist China in 1959," *K'o-hsueh T-ung-pao* 23 (December 11, 1959): 789–90.

17. Chinese Communist Weather Control Experiments (August 1959).

CHAPTER SIX

1. John Lear, "Research in America: Shepherding the Wind," *Saturday Review*, April 2, 1966, 49–52, on 52.

2. National Science Foundation, *Weather Modification, First Annual Report for the Fiscal Year ended 30 June 1959* (Washington, DC: June 1960).

3. "Lightning and Forest Fires in the Northern Rocky Mountain Region," *BAMS* 7 (1926):

117–18; "Bad Forest Fire Weather Responsible for Heavy Losses in Northwest," *BAMS* 7 (1926): 101–2.

4. E. N. Munns, "Forest Fire Meteorology," *BAMS* 7 (1926): 72–73.

5. "Fire Weather Appropriations," *BAMS* 7 (1926): 54.

6. G. S. Lindgren, "Fire Weather in the Adirondacks," *BAMS* 7 (1926): 30–31.

7. Leonard M. Tarr, "Forest Fire Warnings in Connecticut," *BAMS* 9 (1927): 117–19.

8. "Fire Hazard Investigations at the Northeastern Forest Experiment Station," *BAMS* 8 (1927): 85.

9. S. P. Fergusson, "Apparatus for the Forest-Fire Service and Other Projects," *BAMS* 8 (1927): 94–95.

10. George W. Alexander, "Fire Weather and Fire Climate," *BAMS* 11 (1930): 214–15.

11. Vincent J. Schaefer, "The Possibilities of Modifying Lightning Storms in the Northern Rockies," USDA USFS Station Paper No. 18 (Missoula, MT: U.S. Northern Rocky Mountain Forest and Range Experiment Station, 1949).

12. Ibid.

13. Vincent J. Schaefer, "A Proposed Program for the Munitalp Foundation," circa 1951, Schaefer.

14. Vernon Crudge to Lewis Douglas, February 9, 1953, Schaefer.

15. Donald M. Fuquay, "Project Skyfire Progress Report, 1958–1960" (Ogden, UT: Intermountain Forest and Range Experiment Station, USFS, USDA, 1962), 1. By comparison, between 2001 and 2013, an average of ten thousand lightning-caused fires occurred, burning 4.1 million acres. Fire suppression costs for all types of forest fires totaled $1.7 billion in 2013. See "National Interagency Fire Center Lightning-Caused Fires," accessed July 13, 2014, http://www.nifc.gov/fireinfo/firinfo_stats_lightning.html.

16. J. S. Barrows et al., "Project Skyfire," *Final Report of the Advisory Committee on Weather Control, Vol. II* (Washington, DC: Government Printing Office, 1957), 105. See also Vincent J. Schaefer, "Thunderstorms and Project Skyfire," *Transactions of the New York Academy of Science* 17, no. 6, Series II (1955): 470–75.

17. Barrows et al., "Project Skyfire," 108–11. That is, the lightning stemmed from these three types of weather systems. A local air mass is one that has developed in a particular spot, as opposed to the typical use of the term, which includes a substantial area of a continent or an ocean. Frontal system lightning generally forms during a fast-moving cold front, where a colder air mass is replacing a warmer air mass, and the incoming cold air rapidly pushes up the warm air it is replacing, leading to the development of cumulus clouds. High-level, fast-moving lightning storms are generally related to jet-stream movement.

18. Ibid., 112.

19. Ibid., 113.

20. Ibid.

21. Ibid., 115–24.

22. Fuquay, "Project Skyfire," 2–6.

23. National Science Foundation, *Weather Modification—Third Annual Report for Fiscal Year 1961* (Washington, DC: 1962), 20–21.

24. National Science Foundation, *Weather Modification—Fifth Annual Report for Fiscal Year 1963* (Washington, DC: 1964), 22.

25. National Science Foundation, *Weather Modification—Sixth Annual Report for Fiscal Year 1964* (Washington, DC: 1965), 23.

26. National Science Foundation, *Weather Modification—Seventh Annual Report for Fiscal Year 1965* (Washington, DC: 1966), 41. A "long-continuing current" remains in place for at least one millisecond after the initial lightning strike.

27. National Science Foundation, *Weather Modification—Eighth Annual Report for Fiscal Year 1966* (Washington, DC: 1967), 27.

28. James D. Harpster and William J. Douglas, "Weather Modification—A Fire Control Tool," *Journal of Weather Modification* (*JWM*) 3, no. 1 (1971): 244–49.

29. Department of Commerce, *Summary Report: Weather Modification for Fiscal Years 1969, 1970, and 1971* (Washington, DC: US Government Printing Office, 1973), 86–88.

30. US Weather Modification Advisory Board, "The Management of Weather Resources: A Report to the Secretary of Commerce from the Weather Modification Advisory Board (Washington, DC: Weather Modification Advisory Board, US Government Printing Office, 1978), 178.

31. "National Interagency Fire Center Lightning-Caused Fires, 13-year average (2001–2013)," URL: www.nifc.gov/fireinfo/fireinfo_stats_lightng.html, accessed July 13, 2014.

32. M. D. Flannigan and B. M. Wotton, "Climate, Weather, and Area Burned," in *Forest Fires: Behavior and Ecological Effects*, ed. Edward A. Johnson and Kiyoko Miyanishi (San Diego, CA: Academic Press, 2001), 365–69. For more details on lightning's relationship to forest fires, see Don Latham and Earle Williams, "Lightning and Forest Fires," in *Forest Fires,* ed. Johnson and Miyanishi, 376–418.

33. Until the 1980s, BuRec was one of the world's major promoters of reclamation projects. See Paul R. Josephson, *Resources under Regimes* (Cambridge, MA: Harvard University Press, 2004).

34. Jack C. Oppenheimer, NASA, to Archie Goodman, Bureau of Reclamation, April 5, 1961, Box 9/Prof File 61–64 Atmospheric Water Resources Program (AWRP), Dominy. See also, Jack C. Oppenheimer and W. Henry Lambright, "Technology Assessment and Weather Modification," *Southern California Law Review* 45 (1972): 582. While Oppenheimer and Lambright claim that BuRec had to be "pushed into cloud seeding," in an oral history interview, Dominy claimed that he had approached Congress for funding because "the Weather Bureau was not doing a damn thing about investigating whether or not you [man] could make clouds produce more moisture than nature otherwise would provide." See Floyd Dominy oral history interview by Brit Allan Storey, Senior Historian, BuRec, Oral History Program, BuRec, April 6, 1994, and April 8, 1996.

35. Schaefer to Droessler, NSF, Atmospheric Sciences, December 21, 1961, Box 9/Prof File 61–64 AWRP, Dominy.

36. W. U. Garstka and W. C. Munson to Dominy, February 12, 1962; Reclamation's Laboratory in the Sky (Press Release), January 9, 1963, Box 9/Prof File 61–64 AWRP, Dominy; "Cloud Seeding for Rain Called Dakota 'Success,'" AP story in *NYT*, January 9, 1963.

37. B. R. Bellport, January 14, 1963, Box 9/Prof File 61–64 AWRP, Dominy.

38. Assistant to the Commissioner—Information to Dominy, January 25, 1963, Box 9/Prof File 61–64 AWRP, Dominy.

39. "The West—A Potential Future Food Deficit Area," April 4, 1963, Box 13/Prof File 1964 Disasters, Dominy.

40. William I. Palmer to John C. Calhoun, Jr., July 8, 1963, Box 9/Prof File 61–64 AWRP, Dominy.

41. Calhoun to Dominy, July 30, 1963, Box 9/Prof File 61–64 AWRP, Dominy.

42. D. V. McCarthy to Dominy, August 7, 1963, Box 9/Prof File 61–64 AWRP, Dominy. For research on precipitation processes, see, e.g., L. J. Battan, Horace R. Byers, and Roscoe R. Braham, Jr., "The Formation of Precipitation in Convective Clouds," *Proceedings of the Fifth Weather Radar Conference*, Asbury Park, New Jersey (1955).

43. Wendell C. Munson to Chief, Division of Project Development, August 16, 1963, Box 9/Prof File 61–64 AWRP, Dominy.

44. Dominy to Assistant Secretary—Water and Power Development, August 29, 1963; William I. Palmer to Science Adviser, Interior, December 13, 1963; T. W. Mermel to Codes 100, 105, 120, and 700, April 8, 1964, Box 9/Prof File 61–64 AWRP, Dominy.

45. Statement of Senator Clinton P. Anderson, May 21, 1964, Weather Modification, Hearing before the Subcommittee on Irrigation and Reclamation of the Committee on Interior and Insular Affairs, United States Senate, 88th Congress, Second Session, on A Program for Increasing Precipitation in the Colorado River Basin by Artificial Means, 7–8 [hereafter Colorado River Basin hearings].

46. Statements of John C. Calhoun, Jr., May 21, 1964, 10–13, and Walter U. Garstka, May 21, 1964, 32–37, Colorado River Basin hearings.

47. Exchange between John C. Calhoun, Jr., and Senator Clinton P. Anderson, May 21, 1964, Colorado River Basin hearings, 17.

48. Statement of Floyd Dominy, May 21, 1964, Colorado River Basin hearings, 21–23.

49. Senators' comments and questions, May 21, 1964, Colorado River Basin hearings, 29–32.

50. Statement of Walter U. Garstka, May 21, 1964, Colorado River Basin hearings, 32–37.

51. Commissioner's Memorandum No. 166, August 24, 1964, Box 9/Prof File 60–69 AWRP 64–69, Dominy.

52. Department of the Interior/Bureau of Reclamation press release, "Three Major Contracts for Research on Atmospheric Water Resources Awarded by Reclamation," November 11, 1964, Box 56/Press Releases, E/ARM.

53. The Interagency Committee on Atmospheric Sciences (ICAS), established in 1959 by the Federal Council on Science and Technology, shared information and coordinated atmospheric science efforts across the federal government, but did not manage any of its participating agencies and departments. Chief Engineer to Commissioner, Bureau of Reclamation, January 7, 1965, Box 9/Prof File 64–65 AWRP, Dominy.

54. Statement by Advisory Committee on Atmospheric Water Resources to the US Department of the Interior, Bureau of Reclamation, February 5, 1965, Box 9/Prof File 64–65 AWRP, Dominy.

55. Elmer E. Staats to Donald F. Hornig, Leland J. Haworth, Kenneth Holum, and J. Herbert Hollomon, April 23, 1965, Box 9/Prof File 64–65 AWRP, Dominy.

56. A Statement of the Bureau of Reclamation's Strategy for Its Atmospheric Water Resources Program, April 23, 1965, Box 9/Prof File 64–66 AWRP, Dominy.

57. Harry Collins and Robert Evans, *Rethinking Expertise* (Chicago: University of Chicago Press, 2007), 20.

58. Droessler to Garstka, May 13, 1965, Box 9/Prof File 64–66 AWRP, Dominy.

59. Coordinator, Atmospheric Water Resources Program (Archie Goodman) to Commissioner (Dominy), May 20, 1965, Box 9/Prof File 64–66 AWRP, Dominy.

60. Chief Engineer (Barney Bellport) to Commissioner (Dominy), blue envelope letter, June 6, 1965, Box 4/Correspondence 1966 Blue Envelope Letters, Dominy. "Blue envelope letters" were delivered in blue envelopes and carried sensitive correspondence between BuRec officials. Supergrades include General Service (GS) civilian pay grades GS-16 through GS-18.

61. Atmospheric Water Resources Matters for Discussion with Commissioner Dominy, October 5, 1965, Box 9/Prof File AWRP, Dominy. The conferences were the Seventh Interagency Conference on Weather Modification, September 30–October 1, 1965, Shenandoah National Park, Virginia, and the First International Symposium on Water Desalination, October 3–9, 1965, Washington, DC.

62. Coordinator, Atmospheric Water Resources Program, to Commissioner, blue envelope letter, January 21, 1966, Box 9/Prof File 64–66 AWRP, Dominy.

63. Coordinator, Atmospheric Water Resources Program, to Commissioner, blue envelope letter, January 21, 1966, Box 9/Prof File 64–66 AWRP, Dominy.

64. Kenneth Holum to Secretary of the Interior, 18 August 1966, Box 9/Prof File 61–64 AWRP, Dominy.

65. "Plan to Develop Technology for Increasing Water Yield from Atmospheric Sources: An Atmospheric Water Resources Program—US Department of the Interior, Bureau of Reclamation," November 1966, Box 9/Prof File 67–68 AWRP, Dominy.

66. Ibid.; Hearings before the Senate Subcommittee on Water and Power Resources, March 21, 1966, 53–56.

67. Hearings before the Senate Subcommittee on Water and Power Resources, March 22, 1966, 53–64.

68. Ibid.

69. Atmospheric Water Resources Program—Description of Current Efforts, 1967, Box 9/Prof File 67–68 AWRP, Dominy.

70. Dominy to James E. Webb, January 4, 1968, Box 9/Prof File 67–68 AWRP, Dominy.

71. Charles F. Cooper and William C. Jolly, "Ecological Effects of Weather Modification: A Problem Analysis," May 1969, sponsored by Interior/BuRec, Office of Atmospheric Water Resources, Denver Contract No. 14-06-D-6576 of May 1, 1968, University of Michigan School of Natural Resources, Department of Resource Planning and Conservation, Box 121/154, E/AP/REG. On the environmental movement, see, e.g., Adam Rome, *The Bulldozer in the Countryside: Suburban Sprawl and the Rise of American Environmentalism* (Cambridge: Cambridge University Press, 2001); Michael Egan, *Barry Commoner and the Science of Survival: The Remaking of American Environmentalism* (Boston: MIT Press, 2007); and Robert Gottlieb, *Forcing the Spring: The Transformation of the American Environmental Movement* (Washington, DC: Island Press, 1993).

72. Cooper and Jolly, "Ecological Effects."

73. Atmospheric Water Resources Program, Project Skywater, Proceedings, Skywater Conference VI, Numerical Modeling of Atmospheric Processes, US Department of Interior, BuRec, Division of Atmospheric Water Resources Management, Denver, CO, Box 115/Project Skywater, E/AP/RA.

74. Ibid.

75. Ibid. Capshaw and Rader argue that the national security argument worked well to justify projects without "direct military application" such as Skywater. See James H. Capshaw and Karen A. Rader, "Big Science: Price to the Present," in *Science after '40*, ed. Arnold Thackray, *Osiris* 7 (1992): 4.

76. Sidney Shallet, "Electronics to Aid Weather Figuring," *NYT*, January 11, 1946, 12.

77. "Frequently Asked Questions," NOAA Hurricane Research Division, Atlantic Oceanographic and Meteorological Laboratory, accessed February 13, 2016, http://www.aoml.noaa.gov/hrd/tcfaq/C5c.html.

78. Barrington S. Havens et al., "Early History of Cloud Seeding" (Socorro: Langmuir Laboratory, New Mexico Institute of Mining and Technology, 1978), 41–42.

79. "Severe Local Storms for October 1947," *Monthly Weather Review* 75, no. 10 (1947): 203; "Georgia, Carolinas Hit by Hurricane," *NYT*, October 16, 1947, 31.

80. C. P. Mook et al., "An Analysis of the Movement of a Hurricane off the East Coast of the United States, October 12–14, 1947," *Monthly Weather Review* 85 (1957): 243–50.

81. Bernard Vonnegut, "Langmuir's Seeding of Hurricanes," *Science* 169, no. 3940 (July 3, 1970): 8; Irving Langmuir, "Outline of Progress in the Evaluation of Cloud Modification Techniques," 1948, p. 19, Box 93/1948, Langmuir. Langmuir's "Outline" exemplifies how his weather control work met his own criteria for "pathological science," which he defined as science conducted in accordance with acceptable scientific methodologies, but tainted by the investigator's unconscious bias. See Irving Langmuir, "Pathological Science: Colloquium at the Knolls Research Laboratory," December 18, 1953; and James Rodger Fleming, "The Pathological History of Weather and Climate Modification: Three Cycles of Promise and Hype," *Historical Studies in the Physical and Biological Sciences*, 37, no. 1 (2006): 3–25.

82. H. E. Willoughby et al., "Project STORMFURY: A Scientific Chronicle," *BAMS* 66, no. 5 (1985): 505.

83. US Department of Commerce/ESSA, "Project Stormfury: ESSA Fact Sheet," August 1966, 1. For a brief history of the National Hurricane Research Project, see Neal M. Dorst, "The National Hurricane Project: 50 Years of Research, Rough Rides, and Name Changes," *BAMS* 88, no. 10 (2007): 1566–1588.

84. Willoughby et al., "STORMFURY," 506.

85. Pierre St. Amand et al., "Pyrotechnic Production of Nucleants for Cloud Modification, Part I. General Principles," *JWM* 2, no. 2 (May 1970): 25–29.

86. Willoughby et al., "STORMFURY," 506. The lower a hurricane's central barometric pressure, the more powerful it is.

87. J. S. Malkus and R. H. Simpson, "Modification Experiments on Tropical Cumulus Clouds," *Science* 145 (1964): 541–48.

88. J. S. Simpson et al., "1965: Experimental cumulus dynamics," *Reviews of Geophysics* 3 (1965): 387–431.

89. Willoughby et al., "STORMFURY," 507.

90. Ibid.

91. Ibid.

92. Ibid., 508.

93. Ibid., 513.

CHAPTER SEVEN

1. Pierre Saint-Amand, Weather Modification, Hearing of the Senate Committee on Commerce, Las Vegas, Nevada, November 5, 1965, 33.

2. Ibid., 34–36.

3. Ibid., 38.

4. Howard Wiedemann, "Foreign Policy Implications of Weather Modification," attached to Thomas L. Hughes to Secretary of State (Rusk), April 14, 1966, Box 21, State3008D.

5. Ibid.

6. George Perkovich, *India's Nuclear Bomb* (Berkeley: University of California Press, 1999), 14. Nehru would have been right at home with America's Progressive Era leaders.

7. Ibid., 13, 16.

8. Ibid., 25.

9. Barbara D. Metcalf and Thomas R. Metcalf, *A Concise History of India* (Cambridge: Cambridge University Press, 2002), 243.

10. Itty Abraham, *The Making of the Indian Atomic Bomb: Science, Secrecy and the Postcolonial State* (London: Zed Books, 1998), 126.

11. Hans J. Morgenthau, *A New Foreign Policy for the United States* (New York: Frederick A. Praeger, Publishers for the Council on Foreign Relations, 1969), 93. On the use of science and technology in international diplomacy, particularly related to decolonized nations, see Yaron Ezrahi, *The Descent of Icarus: Science and the Transformation of Contemporary Democracy* (Cambridge, MA: Harvard University Press, 1990); and Clark A. Miller, "Science and Democracy in a Globalizing World: Challenges for American Foreign Policy," *Science and Public Policy* 32, no. 3 (2005): 174–86.

12. Abraham, *Making of the Indian Atomic Bomb*, 126.

13. H. W. Brand, *India and the United States: The Cold Peace* (Boston: Twayne Publishers, 1990), 116. On Food for Peace and techno-science's influence on Third World agriculture, see Jacqueline McGlade, "More a Plowshare than a Sword: The Legacy of Cold War Agricultural Diplomacy," *Agricultural History* 83, no. 1 (2009): 79–102; and Lawrence Busch, *The Eclipse of Morality: Science, State and Market* (New York: Aldine de Gruyter, 2000).

14. Dennis Kux, *India and the United States: Estranged Democracies* (Washington, DC: National Defense University Press, 1993), 243.

15. Paul Y. Hammond, *LBJ and the Presidential Management of Foreign Relations* (Austin: University of Texas Press, 1992), 226.

16. Carleton S. Coon Jr. to Carol Laise, March 2, 1966, Box 15/Unlabeled folder, State5255; Brand, *India and the United States*, 118.

17. Lyndon B. Johnson: "Joint Statement Following Discussions with Prime Minister Gandhi of India," March 29, 1966. Online by Gerhard Peters and John T. Woolley, *The American Presidency Project*. http://www.presidency.ucsb.edu/ws/?pid=27518. Accessed February 13, 2016.

18. Orville Freeman to Lyndon B. Johnson, March 22, 1966, CO 113, Box 38/CO 121 India 3/19/66–3/29/66, LBJ/WHCF.

19. Lyndon Baines Johnson, *The Vantage Point: Perspectives of the Presidency, 1963–1969*

(New York: Holt, Rinehart, & Winston, 1971), 226; W. E. Gaithright to Garthoff, Schneider, Coon, and Weiler, July 1, 1966, Box 15/Def 18-2, State5255.

20. Raymond A. Hare to Herman Pollack, September 7, 1966, Box 15/Def 18-2, State 5255.

21. J. Wallace Joyce to Hare, October 10, 1966, Box 17, State3008D.

22. Hammond, *Presidential Management*, 227.

23. E. M. Frisby, "Weather Modification in Southeast Asia, 1966–1972," *JWM* 14 (1982): 1.

24. Ibid., 2.

25. Ibid.

26. Ibid.

27. Ibid., 3.

28. McNamara to Chester Bowles, 091624Z DEC 1966, Box 131/India Memos and Misc., 1 of 2, Vol. VIII 9/66—2/67 [India Memos], LBJ/India. The total cost, exclusive of government salaries: $300,000 for three seeding aircraft, a US Navy weather reconnaissance aircraft, and seventeen people.

29. Ibid. The Australian's name does not appear, but was most likely Edward G. "Taffy" Bowen of the Commonwealth Scientific and Industrial Research Organisation (CSIRO).

30. Rouleau to Joyce, memorandum, December 14, 1966, Box 21, State3008D.

31. Rouleau to Pollack via Joyce, December 16, 1966, Box 21, State3008D.

32. Pollack to Bowles, December 17, 1966, Box 131/India Memos, LBJ/India.

33. Donald Hornig to Walt Rostow, December 22, 1966, Box 131/India Memos, LBJ/India.

34. Rostow to Johnson, December 29, 1966, Box 131/India Memos, LBJ/India.

35. State/Defense to Bowles, December 29, 1966, Box 131/India Memos, LBJ/India.

36. State/Defense to Bowles, December 30, 1966; Bowles to State, 196701061301Z, Box 131/India Memos, LBJ/India.

37. George A. Brown, "Project GROMMET," *Flying*, December 1974, 68–69. My thanks to George Brown's son, Michael, who contacted me and provided a copy of this article. Because of the GROMMET spelling, I would not have found it.

38. Ibid., 69.

39. Ibid., 70.

40. AMEMBASSY New Delhi to NOTS China Lake, US DDR&E [Department of Defense Research and Engineering], Naval Air Systems Command, Naval Weather Service, SecState, 270425Z January 1967, Box 131/India Memos, LBJ/India.

41. AMEMBASSY New Delhi to State, 240512Z January 1967; State/Defense to AMEMBASSY New Delhi, message #127230, January 27, 1967; AMEMBASSY New Delhi to State, 301352Z January 1967; AMEMBASSY New Delhi to State 301340Z January 1967; State/Defense to AMEMBASSY New Delhi, message #129664, February 1, 1967; Howard Wriggins to Rostow, February 1, 1967; AMEMBASSY New Delhi to State, 030632Z February 1967, Box 131/India Memos, LBJ/India.

42. AMEMBASSY to State, 071255Z February 1967, Box 131/India Memos, LBJ/India.

43. Ibid.

44. Brown, "Project GROMMET," 70.

45. State/Defense to AMEMBASSY New Delhi and AMEMBASSY Rawalpindi, message # 133731, February 8, 1967, Box 131/India Memos, LBJ/India.

46. AMEMBASSY New Delhi to State 091256Z February 1967, Box 131/India Memos, LBJ/India.

47. State to AMEMBASSY New Delhi, message #135310, February 10, 1967, Box 131/India Memos, LBJ/India.

48. AMEMBASSY New Delhi to State, 131300Z February 1967, Box 131/India Memos, LBJ/India.

49. AMEMBASSY New Delhi to State, 141231Z February 1967, Box 131/India Memos, LBJ/India.

50. State to AMEMBASSY New Delhi, February 14, 1967, Box 131/India Memos, LBJ/India. This message was cleared by eight people in eight different agencies (including ESSA and Interior) before Acting Secretary of State Katzenbach released it.

51. Brown, "Project GROMMET," 71. Perhaps this is why *Flying* placed Brown's article in the magazine's "sport flying" section.

52. AMEMBASSY New Delhi to State, 201254 February 1967; Wriggins to Rostow, February 21, 1967, Box 131/India Memos, LBJ/India.

53. AMEMBASSY New Delhi to State, 270916Z February 1967, Box 131/India Memos, LBJ/India.

54. AMEMBASSY New Delhi to State, 281256Z February 1967, Box 131/India Memos, LBJ/India.

55. Rostow to Johnson, February 28, 1967, Box 131/India Memos, LBJ/India.

56. P. Koteswaram, *Water from Weather* (Waltair, India: Andhra University Press, 1976); N. Seshagiri, *The Weather Weapon* (New Delhi: National Book Trust, 1977).

57. Walter P. McConaughy quoted in Robert J. McMahon, "Toward Disillusionment and Disengagement in South Asia," in *Lyndon Johnson Confronts the World: American Foreign Policy, 1963–1968*, ed. Warren I. Cohen and Nancy Bernkopf Tucker (Cambridge: Cambridge University Press, 1994), 1–8, on 1.

58. State to AMEMBASSY New Delhi and Rawalpindi, March 8, 1967, Box 131/India Cables, LBJ/India.

59. State to AMEMBASSY New Delhi and Rawalpindi, March 15, 1967, Box 131/India Cables, LBJ/India.

60. Bowles to Hornig, May 11, 1967, Box 131/India Cables, LBJ/India.

61. Hornig to Johnson, June 5, 1967, Box 85/SC Sciences, LBJ/Classified.

62. Hornig to Johnson, February 20, 1967, Box 41/Vietnam Memos, LBJ/NSF.

63. Ibid.

64. Ibid. For a discussion of the environmental impact of weapons, see Richard P. Tucker, "The Impact of Warfare on the Natural World: A Historical Survey," in *Natural Enemy, Natural Ally: Toward an Environmental History of War*, ed. Richard P. Tucker and Edmund Russell (Corvallis: Oregon State University Press, 2004), 15–41.

65. Frisby, "Weather Modification," 3.

66. Rostow to Johnson, May 22, 1967, Box 88/Project Compatriot 5/67–7/67, LBJ/NSF. Michael Adas, *Dominance by Design: Technological Imperatives and America's Civilizing Mission* (Cambridge, MA: Belknap Press, 2006), discusses Rostow's ideas about science and technology as central to modernization, and how he viewed nations such as Vietnam as prime targets for testing new warfare strategies (295).

67. Rostow to Johnson, May 22, 1967, Box 88/Project Compatriot 5/67–7/67, LBJ/NSF.

68. Ibid.

69. Ibid.

70. Ibid.

71. Rostow to Hornig, May 23, 1967; Hornig to Rostow, May 23, 1967, Box 88/Project Compatriot 5/67–7/67, LBJ/NSF.

72. Robert G. Fleagle, *Eyewitness: Evolution of the Atmospheric Sciences* (Boston: American Meteorological Society, 2001), 76.

73. Frisby, "Weather Modification," 3.

74. Rostow to Johnson, June 9, 1967; Johnson to Rostow, June 9, 1967, Box 88/Project Compatriot 5/67–7/67, LBJ/NSF.

75. White House Daily Diary, June 13, 1967, LBJ, accessed February 13, 2016, http://www.lbjlibrary.net/collections/daily-dairy.html.

76. Rostow to Johnson, July 10, 1967, Box 88/Project Compatriot 5/67–7/67, LBJ/NSF. This same file contains Compatriot reports numbered 9–12 (start May 12, 1967) and 14–17 (end July 7–13, 1967). Compatriot reports numbered 5 and 6 are in the National Security File, Subject File Addendum, Box 52/Project Compatriot.

77. Testimony of Lieutenant Colonel Ed Soyster, Office of the Joint Chiefs of Staff, March 20, 1974, Weather Modification Hearings before the Senate Subcommittee on Oceans and International Environment of the Committee on Foreign Relations, 95-10 [Weather Mod Hearings].

78. Dennis J. Doolin, March 20, 1974, 108, Weather Mod Hearings.

79. Soyster, 103. Paul N. Edwards discusses the use of sensors to track movement on the Ho Chi Minh trail—another technologically heavy plan that did not work—in his *Closed World: Computers and the Politics of Discourse in Cold War America* (Boston: MIT Press, 1997), chap. 4.

80. Pierre Saint-Amand, March 20, 1974, 32–37, Weather Mod Hearings. For more on GROMET II in the Philippines, see P. St. Amand, Captain D. W. Reed, USAR, T. L. Wright, and S. D. Elliott, "GROMET II: Rainfall Augmentation in the Philippine Islands," NWC TP 5097, Naval Weapons Center, China Lake, CA, May 1971.

81. March 20, 1974, Weather Mod Hearings 109.

82. March 20, 1974, Weather Mod Hearings, 112.

83. Journal—Office of Legislative Council, June 27, 1972, CIA-RDP74B00415R000300130011-9, CREST.

84. Rear Admiral L. W. Moffit to Commander, Naval Intelligence Command, January 5, 1973, CREST.

85. C. D. Everhart to Chief of Reconnaissance Group Information Requirements Staff, Office of the Deputy Director for Intelligence, CIA, January 17, 1973, CREST.

86. Deputy for Operations, OSA, to Chief, Collection Branch Reconnaissance Group, IRS, February 13, 1973, CREST.

87. Journal—Office of Legislative Council, January 20, 1976, CREST.

88. Prohibiting Hostile Use of Environmental Modification Techniques, Hearing before the Subcommittee on Oceans and International Environment of the Committee on Foreign Relations of the United States Senate, January 21, 1976, 19–23. See Jacob Darwin Hamblin, *Arming Mother Nature: The Birth of Catastrophic Environmentalism* (Oxford: Oxford Univer-

sity Press, 2013), for a full discussion of the ways in which nature was used as a weapon during the Cold War.

89. Claiborne Pell to George H. W. Bush, July 21, 1976, CREST.

90. "U.S. Denies It Tried to Alter Cuba Weather," *Providence Journal*, June 29, 1976.

91. "CIA: Fidel Castro's Helpers," letter to the editor, *NYT*, July 7, 1976, with annotations, CREST.

92. B. C. Evans to Bush, July 16, 1976, CREST.

93. G. A. Carver, Jr., draft letter Bush to Pell, August 5, 1976, CREST.

94. Carver to Bush, August 5, 1976, CREST.

95. Bush to Pell, August 16, 1976, CREST.

CONCLUSION TO PART THREE

1. "Prohibiting Military Weather Modification," Hearings before the Subcommittee on Oceans and International Relations, United States Senate, 92nd Congress, July 26–27, 1972.

2. Stanley A. Changnon, Jr., and W. Henry Lambright, "The Rise and Fall of Federal Weather Modification Policy," *JWM* 19, no. 1 (1987): 5.

3. Barry B. Coble, "Benign Weather Modification," thesis, The School of Advanced Airpower Studies, Maxwell Air Force Base, Alabama, 1996, chaps. 3 and 4. For a view of nonbenign military weather modification possibilities, see Col. Tamzy J. House et al., "Weather as a Force Multiplier: Owning the Weather in 2025," a paper presented to Air Force 2025, Air War College, August 1996.

4. Changnon and Lambright, "Rise and Fall," 1.

5. Stanley A. Changnon, Jr., "The Paradox of Planned Weather Modification," *BAMS* 56, no. 1 (1975): 27.

6. Changnon and Lambright, "Rise and Fall," 4.

7. Ibid.

8. Ibid., 4–5.

9. Ibid., 6.

CONCLUSION

1. Writings on the various aspects of the state are voluminous. Examples of those listed here include: on the administrative state, Stephen Skowronek, *Building a New American State: The Expansion of National Administrative Capacities, 1877–1920* (Cambridge: Cambridge University Press, 1982); on the industrial state, James Gilbert, *Designing the Industrial State: The Intellectual Pursuit of Collectivism in America, 1880–1940* (Chicago: Quadrangle Books, 1972); of the voluminous work on the welfare state, one example is Walter I. Trattner, *From Poor Law to Welfare State: A History of Social Welfare in America,* 6th edition (New York: Free Press, 1998); on the national security state, Bartholomew H. Sparrow, "American Political Development, State-Building, and the 'Security State': Reviving a Research Agenda," *Polity* 40, no. 3 (2008): 355–67; on the warfare state, James T. Sparrow, *Warfare State: World War II Americans and the Age of Big Business* (Oxford: Oxford University Press, 2011); on the proministrative state, Brian Balogh, *Chain Reaction: Expert Debate and Public Participation in American Commer-*

cial Nuclear Power, 1945–1975 (Cambridge: Cambridge University Press, 1991); on the associational state, Brian Balogh, *The Associational State: American Governance in the Twentieth Century* (Philadelphia: University of Pennsylvania Press, 2015); on the scientific state, Jurgen Schmandt and James Everett Katz, "The Scientific State: A Theory with Hypotheses," *Science, Technology, and Human Values* 11, no. 1 (1986): 40–52; and on the gendered nature of the state, Margot Canaday, *The Straight State: Sexuality and Citizenship in Twentieth-Century America* (Princeton, NJ: Princeton University Press, 2009).

2. On "tinkering," see Nancy Langston, *Forest Dreams, Forest Nightmares: The Paradox of Old Growth in the Inland West* (Seattle: University of Washington Press, 1995), 273; on the importation of knowledge from Europe, Daniel P. Rodgers, *Atlantic Crossings: Social Politics in a Progressive Age* (Cambridge, MA: Belknap Press of Harvard University Press, 1998); on the control of nature for military and other purposes, Paul R. Josephson, *Resources under Regimes* (Cambridge, MA: Harvard University Press, 2004), 199; on experts, expertise and the state, Wiebe E. Bijker, Roland Bal, and Ruud Hendriks, *The Paradox of Scientific Authority: The Role of Scientific Advice in Democracies* (Cambridge, MA: MIT Press, 2009); Yaron Ezrahi, *The Descent of Icarus: Science and the Transformation of Contemporary Democracy* (Cambridge, MA: Harvard University Press, 1990); Mark R. Nemec, *Ivory Towers and Nationalistic Minds: Universities, Leadership, and the Development of the American State* (Ann Arbor: University of Michigan Press, 2006); Balogh, *Associational State,* and *Chain Reaction.*

3. Edward Tenner, *Why Things Bite Back: Technology and the Revenge of Unintended Consequences* (New York: Alfred A. Knopf, 1996).

4. Scholars who have addressed "global problems that require global solutions" concerns include Clark A. Miller and Paul N. Edwards, "Introduction: The Globalization of Climate Science and Climate Politics," in *Changing the Atmosphere: Expert Knowledge and Environmental Governance,* ed. Clark A. Miller and Paul N. Edwards, 1–30 (Cambridge, MA: MIT Press, 2001); Josephson, *Resources under Regimes;* Clive Hamilton, *Earthmasters: The Dawn of the Age of Climate Engineering* (New Haven, CT: Yale University Press, 2013); and Clark A. Miller, "Climate Science and the Making of a Global Political Order," in *States of Knowledge: The Co-production of Science and Social Order,* ed. Sheila Jasanoff, 46–66 (London: Routledge, 2007), 48–49.

5. While the overall upward trajectory of the average measured global temperature has become highly politicized, the surrogate natural changes listed here are readily observable. See, for example, Johannes Oerlemans, "Quantifying Global Warming from the Retreat of Glaciers," *Science* 264, no. 5156 (1994): 243–44; Wilfried Thuiller, Sandra Lavorel, Miguel B. Araújo, Martin T. Sykes, and I. Colin Prentice, "Climate Change Threats to Plant Diversity in Europe," *Proceedings of the National Academy of Sciences of the United States of America* 102, no. 23 (2005): 8245–8250; Cecile Cabanes, Anny Cazenave, and Christian Le Provost, "Sea Level Rise during Past 40 Years Determined from Satellite and In Situ Observations," *Science* 294, no. 5543 (2001): 840–42; S. Gruber and W. Haeberli, "Permafrost in Steep Bedrock Slopes and Its Temperature-related Destabilization Following Climate Change," *Journal of Geophysical Research: Earth Surface (2003–2012)* 112, no. F2 (2007). On the global effort to quantify climate change, see Paul N. Edwards, *A Vast Machine: Computer Models, Climate Data, and the Politics of Global Warming* (Cambridge, MA: MIT Press, 2010), chap. 15.

6. See Theda Skocpol, "Bringing the State Back In: Strategies of Analysis in Current Re-

search," in *Bringing the State Back In*, ed. Peter B. Evans, Dietrich Rueschemeyer, and Theda Skocpol, 3–37 (Cambridge: Cambridge University Press, 1985). On the subordination of science to the state, and hence scientists' need for state funding, see Stanley Aronowitz, *Science as Power: Discourse and Ideology in Modern Society* (Houndmills, UK: Macmillan, 1988). Patrick Carroll, *Science, Culture, and Modern State Formation* (Berkeley: University of California Press, 2006), argues "the fundamental objects around which the coproduction of science and government occurs remains constant. They are land, people, and the built environment" (23). That was certainly the case for weather control. Designer weather creates the ultimate "built environment."

7. Skocpol, "Bringing the State Back In," 9.

8. Ibid., 12.

9. Ibid., 14–15.

10. Ibid., 15.

11. Ibid., 8.

12. Faulkner, *Requiem for a Nun*.

13. But depending on the presidential administration in office and the majority party in Congress, funding has been in short supply for a variety of geophysics-related research, particularly related to climate change—a frustrating situation for my colleagues at the American Geophysical Union, for example. Writing in the late 1950s, Hunter Dupree held that "given adequate preparation and presentation of programs, Congress has shown itself able and willing to support research." A. Hunter Dupree, *Science in the Federal Government: A History of Policies and Activities* (Baltimore: Johns Hopkins University Press, 1986), 380. During the early days of weather control, congressional committees were leaders in their support of science that was not only unproven, but highly criticized by meteorologists. Today, the scientific reasoning behind reducing carbon emissions tends to be rejected.

14. S. 2170, 108th Congress, 2nd Session, "Weather Modification Research and Technology Authorization Act."

15. "Committee Approves Bill Establishing Weather Modification Program," KWTX-TV News, Waco, Texas, accessed February 13, 2016, http://www.kwtx.com/home/headlines /1985602.html.

16. The Weather Modification Advisory Committee provides advice on granting licenses and permits to those seeking to enhance rainfall with cloud seeding. See the website of the Texas Department of Licensing and Regulation, accessed February 13, 2016, https://www.tdlr .texas.gov/weather/weathermod.htm.

17. 109th Congress, 1st Session, Senate Report 109–202, filed November 18, 2005.

18. S. 1807, 110th Congress, "Weather Mitigation Research and Development Policy Authorization Act of 2007."

19. Steve Tracton, "NOAA Says No to DHS Hurricane Modification," Capitol Weather Gang Blog, *Washington Post*, August 7, 2009, http://www.washingtonpost.com/blogs/capital -weather-gang/post/noaa-says-no-to-dhs-hurricane-modification/2010/12/20/ABzCv3F_blog .html.

20. Geoengineering Parts I, II, and III, House Committee on Science, Space, and Technology, Subcommittee on Energy and Environment, Serial No. 111-62; 111-75; 111-88, November 5, 2009.

21. See, e.g., Kelsi Bracmort and Richard K. Lattanzio, "Geoengineering: Governance and Technology Policy" (Congressional Research Service, November 26, 2013), ProQuest Congressional (CRS-2011-RSI-089).

22. On geoengineering and its potential downsides see, e.g., Hamilton, *Earthmasters*, and James Rodger Fleming, *Fixing the Sky: The Checkered History of Weather and Climate Control* (New York: Columbia University Press, 2010), chap. 8. Brad Allenby argues that "complex systems cannot be controlled," and makes the case for Earth Systems Engineering and Management, which is not geoengineering. See Allenby, "Technology at the Global Scale: Integrative Cognitivism and Earth Systems Engineering and Management," in *Scientific and Technological Thinking,* ed. Michael E. Gorman, Ryan D. Tweney, David C. Gooding, and Alexandra P. Kincannon, 303–43 (Mahwah, NJ: Lawrence Erlbaum Associates, 2005). Simon Dalby, "Geoengineering: The Next Era of Geopolitics?" *Geography Compass* 9, no. 4 (2014): 190–201, and David W. Keith, "Geoengineering the Climate: History and Prospect," *Annual Review of Energy and Environment* 25 (2000): 245–84, argue that single-state weather control does not count as geoengineering because the scale is not large enough, nor does it count as a "countervailing measure."

23. Hans P. Ahlness, "Airborne Cloud Seeding for Hail Suppression in North America," Reunion International de la WMA, Santiago, Chile, September 25–27, 2013, accessed August 19, 2014, www.weathermodification.org/SantiagoPresentations/Ahlness.pdf.

24. "Utah Division of Water Resources: Cloudseeding," Utah Division of Water Resources, www.water.utah.gov/cloudseeding.

25. Wes Smalling, "In its fourth year, Wyoming's $8.8 million cloud-seeding experiment is drawing big-time attention," February 15, 2009, accessed February 13, 2016, http://trib.com/news/state-and-regional/in-its-fourth-year-wyoming-s-million-cloud-seeding-experiment/article_e0cb5e7-b4dd-52e9-9f46-1051b120e89a.html; "Wyoming, UW Finish Collecting Cloud Seeding Statistical Information," *UW News*, April 4, 2014, accessed February 13, 2016, http://www.uwyo.edu/uw/news/2014/04/wyoming-uw-finish-collection-cloud-seeding-statistical-information.html. Scholarly reports include Bruce A. Boe, James A. Heimbach, Jr., Terrence W. Krauss, Lulin Xue, Xia Chu, and John T. McPartland, "The Dispersion of Silver Iodide Particles from Ground-based Generators over Complex Terrain. Part I: Observations with Acoustic Ice Nucleus Counters," *Journal of Applied Meteorology and Climatology* 53, no. 6 (2014): 1325–1341; and Binod Pokharel, Bart Geerts, and Xiaoqin Jing, "The Impact of Ground-based Glaciogenic Seeding on Orographic Clouds and Precipitation: A Multisensor Case Study," *Journal of Applied Meteorology and Climatology* 53, no. 4 (2014): 890–909.

26. "Weather Modification Program," Colorado Water Conservation Board, accessed February 13, 2016, http://cwcb.state.co.us/water-management/water-projects-programs/Pages/%C2%ADWeather ModificationProgram.aspx.

27. For information on the Weather Modification Association and its mission, see its website: http://www.weathermodification.org/.

28. For a study of conspiracy theories in the United States, see Robert Alan Goldberg, *Enemies Within: The Culture of Conspiracy in Modern America* (New Haven, CT: Yale University Press, 2001). For a journalistic take on HAARP, see "Conspiracy Theories Abound as U.S. Military Closes HAARP," NBCNews, May 22, 2014, accessed February 13, 2016, http://www.nbcnews.com/science/weir-science/conspiracy-theories-abound-u-s-military-closes-haarp

-n112576. For a conspiracy theorist's view of HAARP, see Jerry E. Smith, *HAARP: The Ultimate Weapon of the Conspiracy* (Kempton, IL: Adventures Unlimited Press, 1998). There are over three thousand academic articles on HAARP and its associated data. As of September 2015, the most recent is H. Y. Fu, W. A. Scales, P. A. Bernhardt, S. J. Briczinski, M. J. Kosch, A. Senior, M. T. Rietveld, T. K. Yeoman, and J. M. Ruohoniemi, "Stimulated Brillouin Scattering during Electron Gyro-Harmonic Heating at EISCAT," *Annales Geophysicae* 33, no. 8 (2015): 983–90.

29. On inexpensive nuclear power, see Balogh, *Chain Reaction*, 113; on using Edward Teller's plan to use nuclear weapons to create harbors in the Arctic, see Dan O'Neill, *The Firecracker Boys: H-Bombs, Inupiat Eskimos, and the Roots of the Environmental Movement* (New York: Basic Books, 2007); on the downside of using chemicals to control pests, see Rachel Carson, *Silent Spring* (Boston: Houghton Mifflin, 1962); on prediction and control of the environment, see Carolyn Merchant, *Autonomous Nature: Problems of Prediction and Control from Ancient Times to the Scientific Revolution* (London: Routledge, 2015). On the importance of humility in making scientific and technological decisions, see Sheila Jasanoff, "Technologies of Humility," *Nature* 450 (November 1, 2007): 33. Moral and ethical issues were not taken into account either, and scientists are no more experts in those areas than anyone else. See Lawrence Busch, *The Eclipse of Morality: Science, State and Market* (New York: Aldine de Gruyter, 2000).

30. See Ronald E. Doel and Kristine C. Harper, "Prometheus Unleashed: Science as a Diplomatic Weapon in the Lyndon B. Johnson Administration," *Osiris* 21 (2006): 70. Using weather control as a weapon also played a part in changing American attitudes toward technology from being wildly popular in the 1950s, to skeptical in the 1960s, to downright hostile and fearful in the 1970s. See Thomas Parke Hughes, *Changing Attitudes toward American Technology* (New York: Harper & Row, 1975).

31. See, e.g., the latest reports from the National Research Council: *Climate Intervention: Carbon Dioxide Removal and Reliable Sequestration* and *Climate Intervention: Reflecting Sunlight to Cool Earth* (Washington DC: National Academies Press, 2015), as well as an earlier report on weather modification: National Research Council, *Critical Issues in Weather Modification Research* (Washington, DC: National Academies Press, 2003). All these reports encourage more research, but not necessarily deployment of the results. As René Dubos argues in *Science Awake: Science for Man* (New York: Columbia University Press, 1970), touting science and technology as fixes for all kinds of problems can lead people to ask why they should bother taking the steps that would lessen the problem to begin with (132). I concur in the possibility that a potential future fix may keep people from taking present action, not only on greenhouse gas emissions but also on many other factors that affect their personal lives and the broader environment.

32. See, e.g., "Drought Turns Governments to Cloud Seeding," CBSNEWS/AP, December 11, 2009, accessed February 13, 2016, http://www.cbsnews.com/news/drought-turns-governments-to-cloud-seeding/.

33. And as Paul Josephson has pointed out, "What requires more understanding is how we have come to rely increasingly on science and technology to modify nature, yet fail to fathom the dangers of this approach." Paul R. Josephson, *Industrialized Nature: Brute Force Technology and the Transformation of the Natural World* (Washington, DC: Island Press/Shearwater Books, 2002), 11. On values, see, e.g., Dale Jamieson, "Scientific Uncertainty and the Political

Process," *Annals of the American Academy of Political and Social Science* 545 (1996): 35–43; and Dorothy Nelkin, "The Political Impact of Technical Expertise," *Social Studies of Science* 5 (1975): 35–54. As Harry Collins argues, scientists can provide advice, but they cannot make decisions because those belong to others. See H. M. Collins, *Changing Order: Replication and Induction in Scientific Practice* (London: Sage Publications, 1985), 167.

INDEX

Page numbers in italics refer to figures and tables.